T0214495

SpringerBriefs in Mathematics

Series Editors

Nicola Bellomo
Michele Benzi
Palle Jorgensen
Tatsien Li
Roderick Melnik
Otmar Scherzer
Benjamin Steinberg
Lothar Reichel
Yuri Tschinkel
George Yin
Ping Zhang

SpringerBriefs in Mathematics showcases expositions in all areas of mathematics and applied mathematics. Manuscripts presenting new results or a single new result in a classical field, new field, or an emerging topic, applications, or bridges between new results and already published works, are encouraged. The series is intended for mathematicians and applied mathematicians.

More information about this series at http://www.springer.com/series/10030

SBMAC SpringerBriefs

The **SBMAC SpringerBriefs** series publishes relevant contributions in the fields of applied and computational mathematics, mathematics, scientific computing, and related areas. Featuring compact volumes of 50 to 125 pages, the series covers a range of content from professional to academic.

The Sociedade Brasileira de Matemática Aplicada e Computacional (Brazilian Society of Computational and Applied Mathematics, SBMAC) is a professional association focused on computational and industrial applied mathematics. The society is active in furthering the development of mathematics and its applications in scientific, technological, and industrial fields. The SBMAC has helped to develop the applications of mathematics in science, technology, and industry, to encourage the development and implementation of effective methods and mathematical techniques for the benefit of science and technology, and to promote the exchange of ideas and information between the diverse areas of application.

http://www.sbmac.org.br/

Gabriel Ponce • Régis Varão

An Introduction to the Kolmogorov–Bernoulli Equivalence

Springer

Gabriel Ponce
IMECC
University of Campinas - UNICAMP
Campinas
São Paulo, Brazil

Régis Varão
IMECC
University of Campinas - UNICAMP
Campinas
São Paulo, Brazil

ISSN 2191-8198   ISSN 2191-8201   (electronic)
SpringerBriefs in Mathematics
ISBN 978-3-030-27389-7   ISBN 978-3-030-27390-3   (eBook)
https://doi.org/10.1007/978-3-030-27390-3

Mathematics Subject Classification: 37-XX, 28-XX, 46-XX, 37A35, 37C40, 37D30

This Springer imprint is published by the registered company Springer Nature Switzerland AG.
The registered company address is: Gewerbestrasse 11, 6330 Cham, Switzerland

Preface

Dynamical systems is a broad and active research field in mathematics, part of its success has made into the general population by the name of "Chaos Theory" or "The Butterfly Effect." Ergodic theory is a dynamical system from a probabilistic point of view, and from the ergodic theory point of view we may even list systems according to their chaotic behavior. On the bottom of this list are the Ergodic systems, but on the top of this list are Kolmogorov systems and Bernoulli systems (the most chaotic ones). In many situations, Kolmogorov systems are equivalent to Bernoulli systems, and whether or not a Kolmogorov system is a Bernoulli system is a classical problem in ergodic theory.

This equivalence problem is one of the most beautiful chapters in ergodic theory. It has the quality of combining an incredible amount of techniques and concepts of abstract ergodic theory (such as entropy theory and Ornstein theory) and smooth dynamics (such as nonuniformly hyperbolic dynamics and Pesin's theory). On the other hand, what makes this theory so interesting, that is, so many different and profound beautiful tools, might be a barrier for those willing to enter this area of research. There are no books that could guide a graduate dynamical systems student into this research area (i.e., the Kolmogorov–Bernoulli equivalence problem from the smooth dynamical system point of view) although there are plenty of good books on ergodic theory and smooth dynamics. If one tries to study directly the Kolmogorov–Bernoulli problem for nonuniformly hyperbolic systems, one could easily get stuck into many technicalities (e.g., Pesin's theory) and not grasp the main ideas behind the problem.

This book has been written with the primary purpose of filling this gap so that graduate students could feel comfortable with the idea of working on some of the open problems related to the Kolmogorov–Bernoulli equivalence (or nonequivalence) problem. The way we hope we have accomplished this goal in a rather small book is focusing on the main ideas behind the problem. We may say that the most important part of this book is Chap. 3, where we prove in details the Kolmogorov–Bernoulli equivalence in the context of a toy model (linear Anosov diffeomorphism). This will help the reader to understand main ideas. The subsequent chapter deals with the problem in the context of hyperbolic dynamics. In the last chapter, we briefly go through some more general contexts (such as nonuniformly hyperbolic systems) and present some interesting recent ideas in the area. The first two chapters are the introduction and some preliminaries in ergodic theory.

The reader is assumed to have a working knowledge in ergodic theory and hyperbolic dynamics so that we can focus on the Kolmogorov–Bernoulli equivalence problem itself. We hope the reader may find this book very stimulating and feel interested in doing research on the open problems related to the subject.

Campinas, Brazil   Gabriel Ponce
Campinas, Brazil   Régis Varão
September 2019

Acknowledgements

Gabriel Ponce was partially supported by FAPESP (Grant # 2016/05384-0). Régis Varão was partially supported by FAPESP (Grant # 2016/22475-9) and CNPq.

Contents

Acronyms

SB-partition	Smooth boundary partition
VWB	Very Weak Bernoulli
DA	Derived from Anosov
A^c	Complement of the set A
$\limsup_{n\to\infty} A_n$	$\bigcap_{N=0}^{\infty} \bigcup_{n=N}^{\infty} A_n$
$d(x, y)$	Distance between x and y
$\mathrm{diam}(R)$	Diameter of the set R
$B(x, r)$	Open ball of center x and radius r
∂A	Boundary of the set A
S^1	Circle, $\{z \in \mathbb{C} : |z| = 1\}$
$\mathrm{Im}(f)$	Image of f
$\mathbb{T}^n$	n dimensional Torus given by $\mathbb{R}^n/\mathbb{Z}^n$
Σ_d	$\{(x_n)_{n\in\mathbb{Z}} \mid x_n \in \{1, \ldots, d\}\}$
$< x, n >$	Inner product of x and n
L^*	Adjoint of L
χ_A	Characteristic function of the set A
$L^1(\mu)$	Space of μ-integrable functions
$\|\ \|p$	p-norm at $L^p(\mu)$
$\mathcal{M}(X)$	Space of Borel probability measures on X
$\mathrm{supp}\ \mu$	Support of μ
$f_*\mu$	Push-forward of μ by f
$f^+(x)$	$\lim_{n\to\infty} \frac{1}{n} \sum_{i=0}^{n-1} f(T^i(x))$
$d(\alpha)$	Distribution or probability vector associated to a finite partition α
$\bigvee$	Join of partitions
$J(\mathbf{X}, \mathbf{Y})$	Set of joinings of $\mathbf{X} = (X, \mu)$ and $\mathbf{Y} = (Y, \nu)$
$\bar{d}(\alpha, \beta)$	d-bar distance between the partitions α and β
ξ^-	$\bigvee_{n=0}^{\infty} T^{-n}\xi$
$\Pi(\xi)$	$\mathcal{H}\left(\bigwedge_{n=1}^{\infty} T^{-n}\xi^-\right)$
$\mathcal{H}(\xi)$	Measurable hull of ξ
$\pi(T)$	Pinsker partition of T

$h_\mu(T, \eta)$	Metric entropy of the partition η with respect to T
$H_\mu(\alpha)$	Entropy of the partition α
$H_\mu(\alpha|\beta)$	Conditional entropy of α with respect to β
$h_\mu(f, \alpha)$	Entropy of f with respect to α
$h_\mu(f)$	Entropy of f
$T_p M$	Tangent space of M at p
$\mathrm{Jac}(T)$	Jacobian of $T : X \to Y$
$\mathscr{F}^s$	Stable foliation
$\mathscr{F}^u$	Unstable foliation
$\mathscr{F}^c$	Center foliation
$\mathscr{F}^{cu}$	Center unstable foliation
$\mathscr{F}^{cs}$	Center stable foliation
π^s	Stable holonomy map
π^u	Unstable holonomy map
$\lambda(x, v)$	Lyapunov exponent at x in the direction of v

Chapter 1
Introduction

Abstract In this chapter we briefly motivate the main theme of this book: the Kolmogorov–Bernoulli equivalence problem. We introduce the idea behind the ergodic hierarchy of measure preserving transformations and quickly discuss the problem of detecting conditions under which the Kolmogorov property is promoted to the Bernoulli property. In particular the method introduced by Ornstein and Weiss is of particular interest for our context (smooth dynamics).

1.1 General Ergodic Theory

Given a probability measure space $(X, \mathscr{A}, \mu)$ and a measurable transformation $f : X \to X$, recall that f is said to preserve the measure μ if $\mu(f^{-1}(A)) = \mu(A), \forall A \in \mathscr{A}$. Ergodic theory is the study of the long-time behavior of measurable transformations which preserve a given measure in the ambient space. As we are only interested in this context, by a pair (f, μ) we always mean a measurable transformation $f : X \to X$ preserving a probability measure μ in X.

The first main definition on ergodic theory is ergodicity. We say that the pair (f, μ) is ergodic if for any measurable set A such that $f^{-1}(A) \subset A$, then $\mu(A) \in \{0, 1\}$. We may also say that f is ergodic with respect to the μ.

Ergodicity means that from the measure μ point of view, the dynamics cannot be decomposed into distinct "simpler" dynamics. That is why we shall interpret ergodicity as the first degree of chaoticity of a system.

A basic result in ergodic theory is the Poincaré's recurrence theorem which states that if f preserves the probability μ than given a set A of positive measure, then almost every point of A will return to A infinitely often. But this theorem does not give any more information of the frequency of these returns. Notice that if χ_A is the characteristic function of the set A, then the following limit

$$\lim_{n \to \infty} \frac{1}{n} \sum_{i=0}^{n-1} \chi_A(f^i(x)), \tag{1.1}$$

if it exists, would measure the frequency of return of the point x to the set A.

This frequency of return is better understood by one of the most beautiful and fundamental results of ergodic theory, the Birkhoff's ergodic theorem, which states that for any μ-integrable function $\varphi : X \to \mathbb{R}$, then for μ-almost every point $x \in X$ the sequence of time averages of x converges, that is

$$\lim_{n \to \infty} \frac{1}{n} \sum_{i=0}^{n-1} \varphi(f^i(x)) \tag{1.2}$$

exists. If f is ergodic with respect to μ, then not only the limit (1.2) exists for almost every point but it also converges to the average of φ on the whole space, that is,

$$\lim_{n \to \infty} \frac{1}{n} \sum_{i=0}^{n-1} \varphi(f^i(x)) = \int \varphi d\mu.$$

Therefore, for ergodic systems if we take $\varphi = \chi_A$ we get that

$$\lim_{n \to \infty} \frac{1}{n} \sum_{i=0}^{n-1} \chi_A(f^i(x)) = \int \chi_A d\mu = \mu(A),$$

this means that almost every point returns to A with frequency given by the size of A. This is a remarkable result since it is implied by ergodicity alone, which is not a too restrictive assumption. This rich dynamical behavior implied by ergodicity is yet another reason for us to consider ergodicity as the first degree of chaoticity.

1.2   Chaotic Hierarchy

Further analysis of Birkhoff's ergodic theorem shows that we may characterize the ergodic property by how the dynamics mixtures sets along the time: i.e., (f, μ) is ergodic if and only if for measurable subsets $A, B \subset X$ we have

$$\lim_{n \to \infty} \frac{1}{n} \sum_{i=0}^{n-1} \mu(f^{-i}(A) \cap B) = \mu(A) \cdot \mu(B). \tag{1.3}$$

Thus, not only the set B will intersect A in the future but the average of the intersection has measure $\mu(A) \cdot \mu(B)$. Though every ergodic pair (f, μ) satisfies (1.3), the velocity of the convergence may vary. This originates other types of ergodic properties which reflect different categories of chaos in measure. The *mixing* (in the literature it is also called *strong-mixing*) and the *weak-mixing* properties are two of these types of chaos in measure and are, respectively, defined by:

- (f, μ) is mixing if for every pair of measurable subsets $A, B \subset X$ we have

$$\lim_{n \to \infty} \mu(A \cap f^{-n}(B)) = \mu(A)\mu(B);$$

- (f, μ) is weak-mixing if for every pair of measurable subsets $A, B \subset X$ we have

$$\lim_{n \to \infty} \frac{1}{n} \sum_{i=0}^{n-1} |\mu(A \cap f^{-i}(B)) - \mu(A)\mu(B)| = 0.$$

It is not difficult to see that

$$\text{Mixing} \Rightarrow \text{Weak-Mixing} \Rightarrow \text{Ergodicity}$$

and one can give examples of transformations which are mixing but not weak-mixing, and examples which are weak-mixing but not ergodic, showing that these properties are indeed defining different classes of an hierarchy. So far on top of our hierarchy of chaoticity are the mixing systems.

1.3 Kolmogorov and Bernoulli Systems

Examples of mixing transformations are given by the class of the *Bernoulli shifts*. Given $k \in \mathbb{N}$ and the space of bilateral sequences of k numbers $\Sigma_k = \{0, 1, \ldots, k-1\}^{\mathbb{Z}}$ we define the Bernoulli shift $\sigma_k : \Sigma_k \to \Sigma_k$ by

$$\sigma((x_n)_n) = (x_{n+1})_n,$$

that is, σ shifts the sequence x_n one spot to the left. In the space Σ_k one may define a sigma-algebra $\mathscr{B}_k$ generated by the *cylinders*, i.e., generated by sets of the form

$$[n; a_n, \ldots, a_{n+m}] := \{(x_n)_{n\mathbb{Z}} \mid x_n = a_n, \ x_{n+1} = a_{n+1}, \ldots, x_{n+m} = a_{n+m}\},$$

where $n \in \mathbb{Z}$ and $a_n, \ldots, a_{n+m} \in \{0, \ldots, k-1\}$. Now, given a probability vector $p = (p_0, p_1, \ldots, p_{k-1})$ (that is, a k-uple of real numbers whose sum is equal to one), associated to this probability vector one may define a measure ν called a *Bernoulli measure* for σ_k satisfying

$$\nu([n; a_n, \ldots, a_{n+m}]) = p_{a_n} \cdot p_{a_{n+1}} \cdot \ldots \cdot p_{a_{n+m}}.$$

The pair (σ_k, ν) is mixing and consequently weak-mixing and ergodic. The Bernoulli shift is probably the most important elementary transformation in ergodic theory and its properties go beyond the realm of measure theoretical properties. For instance, σ_k admits very interesting topological properties such as *transitivity* (there is a point in Σ whose orbit is dense) when we consider in Σ the topology generated by the cylinders. It is therefore very important to consider not only the Bernoulli shift itself but also the class of all transformations that can be measurably (or topologically) modeled by Bernoulli shifts. Such class is then called the class of the Bernoulli transformations. More precisely given a probability space $(X, \mathscr{A}, \mu)$, a measure preserving transformation $f : X \to X$ is said to be a *Bernoulli transformation*, or that f has the *Bernoulli property*, if there exist $k \in \mathbb{N}$ and a probability vector p such that f is measurably isomorphic to the Bernoulli shift σ_k endowed with the Bernoulli measure induced by p. It is clear that mixing is preserved by measurable conjugacy, therefore every Bernoulli transformation is mixing. A natural question then arises: does the Bernoulli property define another class in the ergodic hierarchy, i.e., does there exist a mixing transformation which is not Bernoulli? The answer to this question is positive, the Bernoulli property is stronger than the mixing property.

Between the Bernoulli property and the mixing property there is still another class of transformations called *Kolmogorov transformations* or *K-systems*. K-systems, or the K-property, is a concept which was introduced by Andrei N. Kolmogorov in 1958 in a four-page article entitled *A new metric invariant of transient dynamical systems and automorphisms in Lebesgue spaces* (see [5]) and we have learned, from [3], that the original motivation for the nomenclature "K-systems" is that it should be an abbreviation for the word "quasi-regular," whose first letter in Russian is "K." The letter "K" was then interpreted as the initial for "Kolmogorov" and now both nomenclatures are commonly used in the theory. The notion of K-systems is a natural notion for random processes and it essentially states that the present becomes asymptotically independent of all sufficiently long past. More precisely, given a probability measure space $(X, \mathscr{A}, \mu)$ and a measure preserving transformation $f : X \to X$, we say that f is a Kolmogorov automorphism (equiv. Kolmogorov transformation, K-system), or that f has the Kolmogorov property, if there exists a finite generating partition ξ of X such that given any $B \in \bigvee_{-\infty}^{+\infty} f^{-k}\xi$ and $\varepsilon > 0$ there exists an $N_0 = N(\varepsilon, B)$ such that for all $N' \geq N \geq N_0$ and ε-almost every atom A of $\bigvee_N^{N'} f^k\xi$ we have

$$\left| \frac{\mu(A \cap B)}{\mu(A)} - \mu(B) \right| \leq \varepsilon.$$

It is easy to show that the Bernoulli property implies the Kolmogorov property and it is also not complicated to show that the Kolmogorov property implies the mixing property. Thus, the Kolmogorov property is in fact situated between Bernoullicity and mixing. Rohlin–Sinai proved that a system is Kolmogorov if and only if it has completely positive entropy, that is, the entropy of any finite partition is positive. This theorem allows us to exhibit examples which are mixing but not Kolmogorov transformations by taking mixing transformations which have zero entropy. One such example is the horocycle system (see [6]). The problem of show-ing that Bernoulli property and Kolmogorov property are not equivalent in general is much more delicate and it was Ornstein who gave the first example of a Kolmogorov but not Bernoulli transformation. The example was, however, not naturally defined and more natural examples were later given by other mathematicians.

All the concepts and results cited above are given in Chap. 2 where we introduce the reader to the basics of abstract ergodic theory paying special attention to the topics which will be needed in the subsequent chapters such as the Kolmogorov property, the Bernoulli shift, and the relations between the ergodic properties given in the ergodic hierarchy showed above.

1.4 Smooth Ergodic Theory and Hyperbolic Structures

The dynamics described above is in the realm of the abstract ergodic theory, but here we are interested in dynamical systems with more structures, hence parallel to the abstract ergodic theory there is the theory of smooth dynamical systems. One of the main objects of study in this branch of dynamical systems is the class of the Anosov diffeomorphisms. A C^1 diffeomorphism $f : M \to M$ on a compact Riemannian manifold M is said to be an *Anosov diffeomorphism* if the whole manifold is a hyperbolic set for f, i.e., if for every $x \in M$ there is a splitting of the tangent space

$$T_x M = E^s(x) \oplus E^u(x)$$

such that Df has a uniformly contracting behavior along the direction E^s and a uniform expanding behavior along the direction E^u. The directions E^s and E^u are called the stable and the unstable directions of f, respectively. One of the most natural and easy examples of Anosov diffeomorphisms are the linear automorphisms induced on $\mathbb{T}^n$, $n \geq 2$, without eigenvalues of norm one.

More precisely, given an $n \times n$ matrix A with integer entries and determinant of modulus 1, the linear transformation $A : \mathbb{R}^n \to \mathbb{R}^n$, $A(x) := A \cdot x$, induces on $\mathbb{T}^n$ a transformation $L_A : \mathbb{T}^n \to \mathbb{T}^n$ simply by

$$L_A(\pi(x)) = \pi(A \cdot x),$$

where $\pi : \mathbb{R}^n \to \mathbb{T}^n$ is the natural quotient map from $\mathbb{R}^n$ to $\mathbb{T}^n$. In the next chapter we prove that L_A is ergodic with respect to the Lebesgue measure of $\mathbb{T}^n$ if and only if A has no eigenvalues of norm one. That is, L_A is ergodic if and only if A is an Anosov map. The proof of this result is classical and it is actually very short and is based on simple facts of harmonic analysis. For a general C^2 volume preserving Anosov diffeomorphism $f : M \to M$, where M is a compact Riemannian manifold, Anosov proved in his seminal work [1] that f is also ergodic. The proof, which is based on the so-called Hopf argument (see [2]), is completely different from the linear case and is extremely more complicated. It is worth to mention at this point that the proof relies on the geometric and measure theoretical properties of the stable and unstable directions associated to the Anosov diffeomorphism. The same argument actually proves that f is a Kolmogorov automorphism.

In 1971, using an approach characteristic from harmonic analysis, Katznelson [4] showed that actually every ergodic linear automorphism of $\mathbb{T}^n$ is Bernoulli, which is much stronger than ergodicity as we have seen above. Two years later Ornstein and Weiss [7] introduced a more geometric way of proving the same result for linear automorphisms of $\mathbb{T}^2$ and, using the same philosophy, proved that geodesic flows on negatively curved manifolds are also Bernoulli. The Ornstein–Weiss argument is presented in Chap. 3 and may be considered as the main theme of this book since understanding it is fundamental to understand all the extensions it implies in the hyperbolic and non-uniformly hyperbolic context. The crucial advantage of the Ornstein–Weiss argument is that it is based on the geometric and measure-theoretic properties associated with the directions (inside the tangent space) which are invariant by the original map. In other words, the argument introduced by Ornstein–Weiss for linear automorphisms of $\mathbb{T}^2$ seems to show that the presence of an expanding and a contracting direction may cause Kolmogorov property to be promoted to Bernoulli property. Having this in mind it is very natural to ask:

Question 1.1 Let M be a compact Riemannian manifold. If $f : M \to M$ is a volume preserving diffeomorphism with some kind of hyperbolic structure, is it true that if f has the Kolmogorov property then it is also has the Bernoulli property?

This is the main question we are concerned with.

1.5   The Goal of This Book

This book is an invitation for students and dynamical systems specialists willing to research or have some understanding of this classical ergodic theory theme: Kolmogorov–Bernoulli equivalence problem in the context of smooth dynamics. Although there are many excellent books on ergodic theory and smooth dynamics, we feel that there is a gap if one tries to understand the Kolmogorov–Bernoulli equivalence problem by directly going into the research articles. Therefore we have challenged ourselves to contribute on reducing this gap. The way we choose to do so is by focusing on the main ideas one should understand when deciding to do research in the area (such as answering Question 1.1). Since the context we are interested in is the realm of smooth dynamical systems, we have taken for granted some deep abstract ergodic theory results which, although crucial to the theory, are on the realm of abstract ergodic theory and their proofs do not contribute for the ideas we are focusing in, the smooth dynamical context. These abstract ergodic theory results are discussed in Sect. 3.1.2 "Very Weak Bernoulli partitions and Ornstein Theorems."

One of the main machineries for dealing with Question 1.1 are the ideas developed by Ornstein and Weiss [7]. Hence, in order to make the ideas as clear as possible we prove in detail on Chap. 3 the Kolmogorov–Bernoulli equivalence property for the toy model of linear Anosov diffeomorphism on the torus. By doing that we have taken out all extra difficulties imposed by some other general contexts. The subsequent chapter (Chap. 4) proves the same result for a broader class of systems, the hyperbolic ones (or Anosov diffeomorphism). Although new difficulties appear related to this new context, the true ideas of proving the equivalence of Kolmogorov and Bernoulli properties for Anosov diffeomorphisms are still the ones presented in the previous chapter. We keep increasing the generality of our context and in the last chapter ("State of the art"), in a survey like approach, we briefly repeat the main arguments shown in the previous chapters to some much broader context (e.g., partially hyperbolic dynamics, smooth completely hyperbolic maps with singularities and non-uniformly hyperbolic dynamics). We do not intend to make a complete survey of the state of the art on the subject, but instead to serve as some sort of a guide to illustrate on how the ideas on this book are really the ones that take the reader to a further step of maturity on the subject.

References

1. Anosov, D.V.: Geodesic flows on closed Riemannian manifolds of negative curvature. Trudy Mat. Inst. Steklov. **90**, 209 (1967)
2. Hopf, E.: Statistik der Lösungen geodätischer Probleme vom unstabilen Typus. II. Math. Ann. **117**, 590–608 (1940)
3. Katok, A.: Fifty years of entropy in dynamics: 1958–2007. J. Mod. Dyn. **1**, 545–596 (2007)
4. Katznelson, Y.: Ergodic automorphisms of T^n are Bernoulli shifts. Isr. J. Math. **10**, 186–195 (1971)
5. Kolmogorov, A.N.: A new metric invariant of transient dynamical systems and automorphisms in Lebesgue spaces. Dokl. Akad. Nauk SSSR **119**, 861–864 (1958)
6. Marcus, B.: The horocycle flow is mixing of all degrees. Invent. Math. **46**, 201–209 (1978)
7. Ornstein, D.S., Weiss, B.: Geodesic flows are Bernoullian. Isr. J. Math. **14**, 184–198 (1973)

Chapter 2
Preliminaries in Ergodic Theory

Abstract In this chapter we situate the context in which we will work along the book and we recall some central theorems in ergodic theory and entropy theory which will be crucial to the development of the results in the subsequent chapters. This chapter has no intention of being an introductory approach to ergodic theory or entropy theory, but to provide an account of results which will be necessary for the subsequent chapters, therefore proofs of the cited results are omitted and can be found in standard ergodic theory books such as Glasner (Ergodic Theory via Joinings. Mathematical Surveys and Monographs, vol. 101. American Mathematical Society, Providence 2003) and Kechris (Classical Descriptive Set Theory. Graduate Texts in Mathematics, vol. 156. Springer, New York, 1995). We start this chapter with Sect. 2.1 by fixing some notation and recalling some standard definitions and results on the existence of invariant measures for a continuous map. In Sect. 2.2 we state the Birkhoff ergodic theorem and recall the definitions of ergodicity and mixing, two of the properties commonly cited in the ergodic hierarchy. In Sect. 2.3 we fix several notations for the operations among partitions, such as join and intersection, of a certain measurable space which will be used all along the book. In Sect. 2.4 we recall the classical Fubbini's theorem and the much more general Rohklin disintegration theorem. As the Fubbini's theorem is enough for the study of the Bernoulli property for the linear automorphisms of $\mathbb{T}^2$, the much more general Rohklin disintegration theorem is crucial to the study of the Kolmogorov and Bernoulli properties in the uniformly and non-uniformly hyperbolic context of Chaps. 4 and 5. Section 2.5 contains some basic definitions and results on Lebesgue spaces and Sect. 2.6 provides an account of results in entropy theory which will be necessary mainly in the development of Chap. 4. Although we assume the reader to have a working knowledge in ergodic theory and entropy theory, since the main goal of the book is to study the relation between the Kolmogorov and the Bernoulli properties, in Sects. 2.7 and 2.8 we provide carefully the definitions of the Bernoulli and the Kolmogorv properties, as well as proofs to some of the specific results

© The Author(s), under exclusive licence to Springer Nature Switzerland AG 2019
G. Ponce, R. Varão, *An Introduction to the Kolmogorov–Bernoulli Equivalence*,
SpringerBriefs in Mathematics, https://doi.org/10.1007/978-3-030-27390-3_2

which are related to these properties. In particular we recall the structure of the
Bernoulli shift more carefully (see Sect. 2.7.1), we prove that Bernoulli automor-
phisms are Kolmogorov and that Kolmogorov automorphisms are mixing (see
Theorems 2.21 and 2.22).

2.1 Measure Preserving Dynamical Systems

A *measurable space* is a pair $(X, \mathscr{A})$ such that X is a set and $\mathscr{A}$ is a σ-*algebra* over
X, i.e., collection of subsets of X that satisfies

1. $\emptyset \in \mathscr{A}$;
2. if $A \in \mathscr{A}$, then $A^c \in \mathscr{A}$;
3. if $A_0, A_1, \ldots \in \Sigma$, then $\bigcup_{i \in \mathbb{N}} A_i \in \mathscr{A}$.

The elements of $\mathscr{A}$ are called *measurable sets*. Let $(X_1, \mathscr{A}_1)$ and $(X_2, \mathscr{A}_2)$ be
measurable spaces. A *measurable function* from X_1 to X_2 is a function $f : X_1 \to$
X_2 such that, for all $A \in \mathscr{A}_2$, $f^{-1}(A) \in \mathscr{A}_1$.

A *measure space* is a triple $(X, \mathscr{A}, \mu)$ where $(X, \mathscr{A})$ is a measurable space and
μ is a *measure* on $(X, \mathscr{A})$. When $(X, \mathscr{A}, \mu)$ is a measure space for which μ is a
probability measure, that is $\mu(X) = 1$, we say that $(X, \mathscr{A}, \mu)$ is a *probability space*.
It is clear that whenever $\mu(X) < \infty$ one can endow X with a probability measure
by taking the measure $\widetilde{\mu} := \frac{\mu}{\mu(X)}$.

Probability spaces are the central spaces in ergodic theory. When, in addition,
the space X is endowed with some topology, it is very useful to work with $(X, \mathscr{B})$
where $\mathscr{B}$ is the *Borel σ-algebra* of X, i.e., the σ-algebra generated by open sets of
X, and with measures μ defined on this σ-algebra. Such measures are called *Borel
measures*.

Given $(X_1, \mathscr{A}_1, \mu_1)$ and $(X_2, \mathscr{A}_2, \mu_2)$ two measure spaces. A *measure-
preserving function* from X_1 to X_2 is a measurable function $f : X_1 \to X_2$
such that, for all $A \in \mathscr{A}_2$,

$$\mu_1(f^{-1}(A)) = \mu_2(A).$$

When $(X, \mathscr{A}, \mu)$ is a measure space and $f : X \to X$ is a measure preserving
function we also say that μ is an f-*invariant measure* or that f *preserves the
measure* μ.

Definition 2.1 A *measure-preserving dynamical system* is a quadruple $(X, \mathscr{A}, \mu, f)$
such that $(X, \mathscr{A}, \mu)$ is a probability space and $f : X \to X$ is a measure-preserving
function. The function f is also called a *transformation*.

Given a topological space X we denote by $\mathscr{M}(X)$ the set of all Borel probability
measures on X. When X is implicit we will use the simplified notation $\mathscr{M}$. Given
$f : X \to X$ a measurable function, denote by $\mathscr{M}_f$ the set of f-invariant in $\mathscr{M}$,
that is,

$$\mathcal{M}_f = \{\mu \in \mathcal{M} : \mu \text{ is } f - \text{invariant}\}.$$

The space $\mathcal{M}$ can be endowed with the finest topology with respect to which the maps

$$I_\phi : \mathcal{M} \to \mathbb{R}$$

$$\eta \mapsto \int \phi \, d\eta, \quad \phi : X \to \mathbb{R} \text{ continuous,}$$

are continuous. It is not difficult to prove that a base for this topology is the family of all sets of the form

$$V_\mu(f_1, \ldots, f_n; \epsilon) = \left\{ \nu \in \mathcal{M}(X) : \left| \int f_i \, \nu - \int f_i \, d\mu \right| \leq \epsilon, \; i = 1, \ldots, n \right\},$$

where $n \in \mathbb{N}$, $f_1, f_2, \ldots, f_n \in C^0(X)$, $\mu \in \mathcal{M}(X)$, $\epsilon > 0$.

Definition 2.2 The topology on $\mathcal{M}$ constructed above is called the *Weak* Topology* or *Weak star topology*.

Let $\{\phi_k \mid k \in \mathbb{N}\}$ be a countable dense subset in the unitary ball $C^0(X)$. The function d given by

$$d : \mathcal{M} \times \mathcal{M} \to \mathbb{R}$$

$$(\eta_1, \eta_2) \mapsto \sum_{k=1}^{\infty} \frac{1}{2^k} \left| \int \phi_k d\eta_1 - \int \phi_k d\eta_2 \right|$$

is a metric in $\mathcal{M}$ and generates the weak-star topology.

Theorem 2.1 *The space $\mathcal{M}$ endowed with the weak-star topology is a compact metric space.*

The set of measurable functions of X acts naturally over $\mathcal{M}$ by push-forward. More precisely, given any measurable function $f : X \to X$, the push-forward of f on $\mathcal{M}$ is the map f_* defined by

$$f_* : \mathcal{M} \to \mathcal{M}$$

$$\mu \mapsto f_* \mu$$

where

$$f_* \mu(E) := \mu(f^{-1}(E)), \quad E \subset X \text{ measurable}.$$

The following proposition is immediate from the definition of the weak-star topology.

Proposition 2.1 *If X is a topological space and $f : X \to X$ is a continuous function, then $f_* : \mathscr{M} \to \mathscr{M}$ is continuous.*

At this point we state one of the prime steps to start the study of measure invariant maps, the Krylov–Bogolubov Theorem.

Theorem 2.2 (Krylov–Bogolubov) *Let $f : X \to X$ be a continuous map on the compact metric space X. Then f admits an invariant probability measure.*

2.2 Birkhoff's Ergodic Theorem and the Ergodic Property

Let $(X, \mathscr{A}, \mu)$ be a measure space and $T : X \to X$ a measurable function. The n-th *Cesàro mean* or *Birkhoff average* for a measurable function $f : X \to \mathbb{R}$ is defined as

$$S_n(f, x) = \frac{1}{n} \sum_{i=0}^{n-1} f(T^i(x)).$$

Theorem 2.3 (Birkhoff Ergodic Theorem) *Let $T : X \to X$ be a measurable maps preserving a probability measure μ. Given any integrable function $f : X \to \mathbb{R}$, the limit*

$$\widetilde{f}(x) = \lim_{n \to \infty} \frac{1}{n} \sum_{j=0}^{n-1} f(T^j(x))$$

exists for μ-almost every point $x \in X$ and, moreover, the function $\widetilde{f}$ defined in this way is integrable and satisfies

$$\int \widetilde{f}(x) d\mu = \int f(x) d\mu.$$

The following is also a useful consequence of Birkhoff's Ergodic Theorem.

Proposition 2.2 *Let $T : X \to X$ be an invertible measurable map preserving a probability measure μ on X. Given any $f \in L^2(\mu)$, for μ-almost every $x \in X$ the limit of Birkhoff averages of f with respect to T coincides with the limit of Birkhoff averages of f with respect to T^{-1}, that is,*

$$\lim_{n \to \infty} \frac{1}{n} \sum_{j=0}^{n-1} f(T^j(x)) = \lim_{n \to \infty} \frac{1}{n} \sum_{j=0}^{n-1} f(T^{-j}(x)), \quad \mu - a.e. \ x \in X.$$

Definition 2.3 Let $(X, \mathscr{A}, \mu)$ be a measure space. A measurable map $f : X \to X$ is said to be ergodic if, for any measurable f-invariant subset $A \subset X$ (i.e, $f(A) = A$), we have $\mu(A) \in \{0, 1\}$. We say that μ is an invariant ergodic measure if μ is f-invariant and ergodic.

All along the book we are working with invariant measures, thus we will refer to invariant ergodic measures simply as ergodic measures. The following well-known theorem provides several characterizations of ergodicity.

Theorem 2.4 *Let $T : X \to X$ be a measurable function preserving a probability measure μ on X. The following are equivalent:*

(a) μ is an invariant ergodic measure;
(b) for all $\phi \in L^1(\mu)$, the function $\widetilde{\phi}$ is constant μ-a.e.;
(c) for all $\phi \in L^1(\mu)$

$$\widetilde{\phi}(x) = \int \phi(y) \, d\mu(y),$$

for μ-almost every $x \in X$;
(d) If $\phi : X \to \mathbb{R}$ is T-invariant almost everywhere in the sense that

$$\phi(T(x)) = \phi(x), \quad \mu - a.e.,$$

then ϕ is constant μ-a.e.;
(e) If $A \subset M$ such that $T^{-1}(A) \subset A$, then $\mu(A) = 0$ or $\mu(A) = 1$;
(f) If $A \subset M$ such that $T^{-1}(A) \supset A$, then $\mu(A) = 0$ or $\mu(A) = 1$.

In the sequel we recall how to prove that linear automorphisms of tori are ergodic if, and only if, none of the eigenvalues is a root of unity. As this class of automorphisms is crucial to this book we do the main part of this example in detail.

Example 2.1 Consider the 2-torus $\mathbb{T}^2 = \mathbb{R}^2/\mathbb{Z}^2$. Given a 2×2 matrix with integer entries and determinant of modulus 1, $L : \mathbb{R}^2 \to \mathbb{R}^2$, since $L \cdot \mathbb{Z} = \mathbb{Z}$ the matrix L naturally induces a map on $\mathbb{T}^2$, which we will denote by $f : \mathbb{T}^2 \to \mathbb{T}^2$, defined by

$$\pi \circ L = f \circ \pi,$$

where $\pi : \mathbb{R}^2 \to \mathbb{T}^2$ is the natural projection. It is not difficult to see that since $|\det L| = 1$, L preserves the standard Lebesgue measure m on $\mathbb{T}^2$. Let us show that L is ergodic if, and only if, L has no eigenvalues of modulus one.

Let $\varphi \in L^2$ an f-invariant function, i.e., $\varphi \circ f = \varphi$ a.e. Consider the Fourier expansion of φ:

$$\varphi(x) = \sum_{n \in \mathbb{Z}^2} a_n \cdot e^{2\pi i <x,n>}.$$

Therefore we have

$$\varphi(f(x)) = \sum_{n \in \mathbb{Z}^2} a_n \cdot e^{2\pi i <L \cdot x, n>} = \sum_{n \in \mathbb{Z}^2} a_n \cdot e^{2\pi i <x, L^* \cdot n>}.$$

Since $\varphi(f(x)) = \varphi(x)$ a.e we must have

$$\sum_{n \in \mathbb{Z}^2} a_n \cdot e^{2\pi i <x, n>} = \sum_{n \in \mathbb{Z}^2} a_n \cdot e^{2\pi i <x, L^* n>}.$$

By the uniqueness of the expansion it follows that

$$a_{L^* n} = a_n$$

for all $n \in \mathbb{Z}^2$. For any fixed $n \in \mathbb{Z}$, the Bessel inequality implies

$$\sum_{i \in \mathbb{Z}} |a_{(L^*)^i n}|^2 \le \sum_{n \in \mathbb{Z}^2} |a_n|^2 = \|\varphi\|_2^2 < +\infty.$$

Consequently, for any $n \in \mathbb{Z}^2 \setminus \{(0, 0)\}$ either $a_n = 0$ or there exists $i \in \mathbb{Z}$ such that

$$(L^*)^i n = n.$$

The latter occurs if, and only if, L has unitary eigenvalues. Thus, if L has no unitary eigenvalues we have $\varphi(x) = a_{(0,0)}$ which implies the ergodicity of f. The other direction (if L has unitary eigenvalues then f admits a non-constant invariant function) is simpler and is left as an exercise for the reader.

Ergodic measures have a special place inside the set of all invariant probability measures of a given measurable function. To state precisely this fact let us recall two basic definitions.

If X is a convex set, we say that $p \in X$ is an *extremal point of X* if p cannot be written as a convex combination of other two points in X, that is,

if $p = tx + (1 - t)y$ with $x, y \in X$ and $t \in [0, 1]$, then $x = p$ or $y = p$.

Given two measures μ and ν on a measure space X, ν is said to be *absolutely continuous* with respect to μ, and denoted as $\nu << \mu$, when

$$A \subset X \text{ measurable}, \ \mu(A) = 0 \Rightarrow \nu(A) = 0.$$

Proposition 2.3 *Let $f : X \to X$ be a measurable function on a measurable space X.*

(a) *Let ν be an f-invariant probability measure and μ be an f-invariant ergodic probability measure. If $\nu << \mu$, then $\nu = \mu$.*

(b) *Let μ be an f-invariant probability measure. Then*

$$\mu \text{ is ergodic } \Leftrightarrow \mu \text{ is an extremal point of } \mathcal{M}.$$

As a consequence of this characterization one can prove the existence of invariant ergodic probability measures for any continuous function on metric spaces.

Theorem 2.5 *Let $f : X \to X$ be a continuous function on a compact metric space. Then f admits a Borel ergodic probability measure.*

Other classical properties in the hierarchy of ergodic properties are the mixing and the weak-mixing properties.

Let $f : X \to X$ be a measurable map preserving a probability measure μ on X. The *mixing* (in the literature it is also called *stong-mixing*) and the *weak-mixing* properties are, respectively, defined by:

- (f, μ) is mixing if for every pair of measurable subsets $A, B \subset X$ we have

$$\lim_{n \to \infty} \mu(A \cap f^{-n}(B)) = \mu(A)\mu(B);$$

- (f, μ) is weak-mixing if for every pair of measurable subsets $A, B \subset X$ we have

$$\lim_{n \to \infty} \frac{1}{n} \sum_{i=0}^{n-1} |\mu(A \cap f^{-i}(B)) - \mu(A)\mu(B)| = 0.$$

It is easy to see that

$$\text{Mixing} \Rightarrow \text{Weak-Mixing} \Rightarrow \text{Ergodicity.}$$

Later we will position the Bernoulli and the Kolmogorov properties in this hierarchy as well.

2.3 Operations with Partitions

A partition α of a measure space $(X, \mathscr{A}, \mu)$ is a collection of measurable subsets of X, called atoms of α, satisfying:

1. $A \cap B = \emptyset$ for every pair of distinct atoms $A, B \in \alpha$;
2. $\bigcup_{A \in \alpha} A = X$.

The sets which are elements of a partition α are also called *atoms* of α.

Given two partitions α and β of X, we say that α *refines* β, or that it is *finer* than β or even that β is coarser than α, and denote it by $\beta < \alpha$ (or equivalently $\alpha > \beta$), if each element of β is a union of element of α except for a set of zero measure. Given any two partitions α and β of X, we define the *join* of α and β to be the smallest partition which refines both α and β. More precisely, the join of α and β, denoted by $\alpha \vee \beta$ is the partition given by

$$\alpha \vee \beta := \{A \cap B : A \in \alpha, B \in \beta\}.$$

If $\{\alpha_i\}_{i=m}^{n}$ is a sequence of partitions of X, we denote by $\bigvee_{m}^{n} \alpha_i$ the *refinement* of all the partitions in the sequence, that is, $\bigvee_{m}^{n} \alpha_i$ is the partition of X which is finer than α_i, for every $m \leq i \leq n$ and, if ξ is a partition of X which is finer than α_i, for every $m \leq i \leq n$, then $\bigvee_{m}^{n} \alpha_i < \xi$.

We say that a sequence of partitions $\{\alpha_i\}_{i=1}^{\infty}$ is an *increasing sequence of partitions converging to* α if $\alpha_1 < \alpha_2 < \ldots$ and $\bigvee_{n=1}^{\infty} \alpha_i = \alpha$.

Given two partitions α and β of X, we define the *intersection* of α and β, denoted by $\alpha \wedge \beta$, to be the finest partition which is coarser than both α and β, that is, $\alpha \wedge \beta < \alpha, \alpha \wedge \beta < \beta$ and if

$$\eta < \alpha \quad \text{and} \quad \eta < \beta$$

then $\eta < \alpha \wedge \beta$.

When referring to equivalence or equality of two σ-algebras in a measure space (X, μ) we always mean equality modulo zero, that is, along this section we will say that two σ-algebras $\mathscr{A}$ and $\mathscr{B}$ are equal if for every element $A \in \mathscr{A}$ there exists an element $B \in \mathscr{B}$ such that $\mu(A \Delta B) = 0$ and vice versa.

Given a measure space $(X, \mathscr{A}, \mu)$ partitions of X are very important tools to code a certain dynamical system $T : X \to X$. It turns out that some codes may be more "precise" than others and this can be concluded by comparing, in some sense, the partitions which generate them. In what follows we will state some basic facts on the structure of the space of partitions with the goal of introducing a metric in this space.

A particularly important type of partitions are the finite partitions. All along the book finite partitions are always assumed to be ordered partitions, that is, by a finite partition $\alpha = \{A_1, \ldots, A_n\}$ of a measure space $(X, \mathscr{A}, \mu)$ we mean an *ordered finite collection of measurable sets which constitute a partition of* X. In case α and β are finite partitions, then the join $\alpha \vee \beta = \{A_i \cap B_j : A_i \in \alpha, B_j \in \beta\}$, which is clearly a finite partition of X, is considered to be ordered according to the lexicographic ordering.

For convenience, if $\{\alpha_i\}_{i=-\infty}^{\infty}$ is an infinite sequence of finite partitions, we will, by abuse of notation, denote by

$$\bigvee_{-\infty}^{\infty} \alpha_i$$

the smallest complete σ-algebra containing $\bigcup_{n=-\infty}^{\infty} \alpha_n$.

Definition 2.4 Let $(X, \mathscr{A}, \mu)$ be a probability space and $T : X \to X$ a measurable automorphism. We say that a partition α of X is a generating partition for T if $\mathscr{A}$ is the smallest complete σ-algebra containing $\bigcup_{n=-\infty}^{\infty} T^n \alpha$, that is, if $\bigvee_{-\infty}^{\infty} \alpha_i = \mathscr{A}$.

2.4 Measure Disintegration

Let $(X, \mathscr{A}, \mu)$ and $(Y, \mathscr{B}, \nu)$ be two (finite) measure spaces and let $(X \times Y, \mathscr{A} \times \mathscr{B}, \mu \times \nu)$ to be the product space. For $E \subset X \times Y$ and $x \in X$, the x-section of E, or the section of E determined by x is the set

$$E_x = \{y \in Y : (x, y) \in E\}.$$

Analogously, for $y \in Y$ we define the y-section of E, or the section of E determined by y, by

$$E^y = \{x \in X : (x, y) \in E\}.$$

Given a measurable function h defined on a measurable subset $E \subset X \times Y$, for each $x \in X$ we define the x-section of h, or the section of h determined by x, to be the function $h_x : E_x \to \mathrm{Im}(h)$ given by

$$h_x(y) := h(x, y).$$

Analogously, for each $y \in Y$ we define the x-section of h, or the section of h determined by y, to be the function $h^y : E^y \to \mathrm{Im}(h)$ given by

$$h^y(x) := h(x, y).$$

The Fubini's Theorem, which is a classical result in measure theory, provides a disintegration of the product measure along the x-sections or along the y-sections.

Theorem 2.6 (Fubini's Theorem, See [3, p. 148, Theorem C]) *Let (X, μ) and (Y, ν) be measure spaces. If h is an integrable function on $X \times Y$, then almost every section of h is integrable. If the functions f and g are defined by*

$$f(x) = \int h(x, y) d\nu(y) \quad and \quad g(y) = \int h(x, y) d\mu(x),$$

then f and g are integrable and

$$\int h d(\mu \times \nu) = \int f d\mu = \int g d\nu.$$

2.4.1 Rokhlin's Disintegration Theorem

Let $(M, \mu, \mathscr{B})$ be a probability space, where M is a compact metric space, μ a probability and $\mathscr{B}$ a Borel σ-álgebra. Given a partition $\mathscr{P}$ of M by measurable sets we define a measure space $(\mathscr{P}, \widetilde{\mu}, \widetilde{\mathscr{B}})$ by the following: let $\pi : M \to \mathscr{P}$ be the canonical projection which associates to a point of M the element of the partition which contains it, we define $\widetilde{\mu} := \pi_* \mu$ e $\widetilde{\mathscr{B}} := \pi_* \mathscr{B}$.

Definition 2.5 Let $\mathscr{P}$ be a partition. A family $\{\mu_P\}_{p \in \mathscr{P}}$ is a *system of conditional measures*, or a *disintegration*, for μ (with respect to $\mathscr{P}$) if

(i) Given $\phi \in C^0(M)$, then $P \mapsto \int \phi \, d\mu_P$ is measurable;
(ii) $\mu_P(P) = 1$ $\widehat{\mu}$-a.e.;
(iii) If $\phi \in C^0(M)$, then $\displaystyle \int_M \phi \, d\mu = \int_{\mathscr{P}} \left(\int_P \phi \, d\mu_P \right) d\widetilde{\mu}$

Observe that the conditions (i) and (iii) are also true for ϕ bounded, by the dominated convergence theorem. When the family is implicit, we may also say that the family $\{\mu_P\}$ *disintegrates* the measure μ.

As the following proposition shows, the system of conditional measures which disintegrates a given measure is unique when restricted to a full measure set of elements of the given partition.

Proposition 2.4 *If $\{\mu_P\}$ and $\{\nu_P\}$ are conditional measures which disintegrate μ, then $\mu_P = \nu_P$ $\widetilde{\mu}$-a.e.*

Proof Suppose, by absurd, that exists $\mathscr{Q} \subset \widetilde{\mathscr{B}}$ with $\widetilde{\mu}(\mathscr{Q}) > 0$ such that $\mu_P \neq \nu_P$ for all $P \in \mathscr{Q}$.

Claim There exists $\mathscr{Q}_0 \subset \mathscr{Q}$ with $\widetilde{\mu}(\mathscr{Q}) > 0$ and $\phi \in C^0(M)$ such that $\int_P d\mu_P > \int_P d\nu_P$ for all $P \in \mathscr{Q}_0$ or $\int_P d\mu_P < \int_P d\nu_P$ for all $P \in \mathscr{Q}_0$.

In fact, let $\{\phi_k\}$ be a dense countable set of $C^0(M)$. Define the sets

$$A_i = \{P \in \mathscr{Q} \mid \int_P d\mu_P \neq \int_P d\nu_P\}.$$

Since $\widetilde{\mu}(\cup_i A_i) = \widetilde{\mu}(Q) > 0$, there exist i_0 such that $\widetilde{\mu}(A_{i_0})$ with $\widetilde{\mu}(Q_{i_0}) > 0$. Now the claim follows.

Let ϕ and $\mathscr{Q}$ as in the claim above.

$$\int \phi \chi_{\pi^{-1}(\mathscr{Q})} d\mu = \int (\int \phi \chi_{\pi^{-1}(\mathscr{Q})} d\mu_P) d\widetilde{\mu}(P) = \int_{\mathscr{Q}} (\int \phi \, d\mu_P) d\widetilde{\mu}(P)$$

$$> \int_{\mathscr{Q}} (\int \phi \, d\nu_P) d\widetilde{\mu}(P) = \int \phi \chi_{\pi^{-1}(\mathscr{Q})} d\mu$$

Which yields an absurd.   □

Corollary 2.1 *If $T : M \to M$ preserves a probability μ, then it preserves μ_P for $\tilde{\mu}$-a.e P.*

Proof We only need to see that $\{T_*\mu_P\}_{P \in \mathscr{P}}$ also disintegrates μ. □

Definition 2.6 We say that $\mathscr{P}$ is a measurable partition if, besides being a partition, there exists a family of Borel sets $\{A_i\}_{i \in \mathbb{N}}$ such that

$$\mathscr{P} = \{A_1, A_1^c\} \vee \{A_2, A_2^c\} \vee \ldots \bmod 0.$$

Theorem 2.7 (Rokhlin Disintegration) *Let $\mathscr{P}$ be a measurable partition for the compact set M and μ a borel probability. Then, there exists a disintegration of μ.*

For a detailed proof of Theorem 2.7 we refer the reader to [2].

2.5 Basics on Lebesgue Spaces

In this section we recall some properties and characterizations of the basic measure theoretic setting in which we will work. The measure spaces in which we are interested in are measure spaces with a topological structure roughly resembling the structure of the unit interval plus a countable number of isolated points. These spaces will be called along the rest of the book as *Lebesgue spaces*. In the literature it is also common to find the nomenclature *standard probability space* or *Lebesgue-Rokhlin probability space*.

Recall that an isomorphism ϕ between two measure spaces $(X_1, \mathscr{A}_1, \mu_1)$ and $(X_2, \mathscr{A}_2, \mu_2)$ is a bijective map $\phi : X_1 \to X_2$ such that ϕ and ϕ^{-1} are measurable and measure preserving. We say that ϕ is an isomorphism (mod 0) is there are sets $\tilde{X}_i \subset X_i$ such that

$$\mu_1(X_1 - \tilde{X}_1) = \mu_2(X_2 - \tilde{X}_2) = 0$$

and ϕ is an isomorphism from $\tilde{X}_1$ to $\tilde{X}_2$ with the, respectively, induced σ-algebras. Also recall that in the case $X_1 = X_2 = X$, if $\phi : X \to X$ is an isomorphism (mod 0) then we say that ϕ is an automorphism of X.

Definition 2.7 We say that a probability space (P, p) is a space of atoms if P is countable and each element of P has positive p-measure, that is, $p(\{x\}) > 0, \forall x \in X$.

Definition 2.8 A probability space $(X, \mathscr{A}, \mu)$ is called a Lebesgue space if it is isomorphic mod 0 to $([0, 1], Leb)$, to a space of atoms (P, p) or to a disjoint union of both. If $(X, \mathscr{A}, \mu)$ is a Lebesgue space without atomic part, we say that X is a non-atomic Lebesgue space.

In what follows we briefly present the characterization of Lebesgue spaces given by Rohlin [8].

Definition 2.9 Let $(X, \mathscr{A}, \mu)$ be a probability space and $\mathscr{E} \subset \mathscr{A}$ a collection of sets. We say that:

(1) $\mathscr{E}$ separates X if there is a set $E \subset X$ with $\mu(E) = 0$ such that for any pair of distinct points $x, y \in X \setminus E$ there exists a set $A \in \mathscr{E}$ such that

$$x \in A, y \notin A \quad \text{or} \quad x \notin A, y \in A.$$

(2) $\mathscr{E}$ generates $\mathscr{A}$ if $\mathscr{A}$ is the smallest complete σ-algebra containing $\mathscr{E}$.
(3) $\mathscr{E}$ is complete if $\mathscr{E} = \{A_n : n \in \mathbb{N}\}$ is a countable collection of sets such that each intersection of the form

$$\bigcap_{n \in \mathbb{N}} A_n^*, \quad A_n^* \in \{A_n, X \setminus A_n\}$$

are non-empty.

For the sake of simplicity we also say that a measure space $(X, \mathscr{A}, \mu)$ is a subspace of another measure space $(Y, \mathscr{B}, \nu)$ if $X \in \mathscr{B}$ and $\mathscr{A}$ and μ are, respectively, the induced σ-algebra and the induced measure on X, that is:

$$\mathscr{A} = \{B \cap X : B \in \mathscr{B}\}, \quad \mu(B \cap X) := \nu(B) \text{ for all } B \in \mathscr{B}.$$

Theorem 2.8 ([8]) *A probability space $(X, \mathscr{A}, \mu)$ is a Lebesgue space if and only if it is a subspace of a probability space $(Y, \mathscr{B}, \nu)$ which has a complete, separating, generating sequence.*

The next two theorems are very useful and we refer the reader to [8] for their proofs.

Theorem 2.9 ([8], See Also Exercise 15.4 **from [4])** *If $(X, \mathscr{A}, \mu)$ is a Lebesgue space and $\mathscr{E} \subset \mathscr{A}$ is a collection of sets. Then $\mathscr{E}$ separates X if and only if $\mathscr{E}$ generates $\mathscr{A}$.*

Theorem 2.10 (See [8, p. 20]) *Let $(X_i, \mathscr{A}_i, \mu_i)$, $i = 1, 2$, be Lebesgue spaces and $\phi : X_1 \to X_2$ be a measurable measure-preserving mapping. If ϕ is injective (mod 0), then ϕ is surjective (mod 0), in particular, $\phi(X_1)$ must be $\mathscr{A}_2$-measurable.*

2.6 Some Results on Entropy Theory

Entropy is one of the most important invariants in dynamical systems whose role is to capture the asymptotic chaos generated by the system. However the only place where entropy will be used along the book is in Sect. 4.2 to give another, equivalent, characterization of the Kolmogorov property. Therefore, here we are far from giving an adequate introduction to the entropy theory and we will restrict ourselves to present only its definition and a few facts. An interested reader may find a more complete account on metric entropy in the book of Walters [11].

Let $\alpha = \{A_i\}_{i=1}^k$ be a finite partition of X by measurable sets and μ a probability measure on X. We define the *entropy* $h_\mu(\alpha)$, of the partition α with respect to the measure μ, by:

$$H_\mu(\alpha) = -\sum_{i=1}^k \mu(A_i) \, \log \, \mu(A_i).$$

Given two finite partitions $\alpha = \{A_i\}_{i=1}^k$ and $\beta = \{B_j\}_{j=1}^l$ of X, we define the *conditional entropy* of α with respect to β by

$$H_\mu(\alpha|\beta) = -\sum_{i,j} \mu(A_i \cap B_j) \log \left(\frac{\mu(A_i \cap B_j)}{\mu(B_j)} \right)$$

The following is a classical fact from entropy theory.

Proposition 2.5 *Let $f : X \to X$ be a measurable map preserving a probability measure μ on X. Given a finite partition ξ of X we have*

$$h_\mu(T, \xi) = \lim_n H_\mu \left(\xi | \bigvee_{j=1}^n f^{-j}\xi \right).$$

Let $f : X \to X$ be a measurable function and $\alpha = \{A_i\}$ a partition. We denote $\alpha^n = \alpha \vee f^{-1}\alpha \vee \ldots \vee f^{-(n-1)}\alpha$. Now, we define the entropy of f with respect to a probability μ and a partition α to be the following limit:

$$h_\mu(f, \alpha) = \lim_n \frac{1}{n} H_\mu(\alpha^n) = \inf_n \frac{1}{n} H_\mu(\alpha^n). \tag{2.1}$$

It is part of the development of the theory to check that the equalities above are correct and that the limit indeed exists, once again we refer the reader to [11] for details.

Definition 2.10 The metric entropy of f with respect to μ, of just the entropy of f with respect to μ, is defined by

$$h_\mu(f) = \sup_\alpha h_\mu(f, \alpha)$$

where the supremum is taken over all finite partitions α of X by measurable sets.

The next theorem, proved by Kolmogorov, shows that entropy can be evaluated by its value on generating partitions.

Theorem 2.11 (See [11]) *If ξ is a generating partition, then*

$$h_\mu(f) = h_\mu(f, \xi).$$

As we will state a fortiori, metric entropy is a complete invariant for the family of Bernoulli automorphisms (see Theorem 2.18).

2.6.1 The Pinsker Partition and Systems with Completely Positive Entropy

Consider $(X, \mathscr{A}, \mu)$ a Lebesgue space and let ξ be a partition of X. Denote by $\mathscr{B}(\xi) \subset \mathscr{A}$ the sub σ-algebra generated by the elements of ξ, that is, $\mathscr{B}(\xi)$ is the family of all measurable subsets of X which are unions of elements of ξ. Let $\mathscr{B} \subset \mathscr{A}$ be a sub σ-algebra of $\mathscr{A}$. Let us construct a partition ξ of X such that $\mathscr{B}(\xi)$ is equal to $\mathscr{B}$ (recall that by equality of σ-algebras we mean equality mod zero). Let Y be the (at most countable) set of atoms of μ which are elements of $\mathscr{B}$. Now, let $X' = X \setminus Y$ and consider $\mathscr{B}'$ and μ' to be the restrictions of $\mathscr{B}$ and μ to X', respectively. As $(X, \mathscr{A}, \mu)$ is a Lebesgue space so are $(X, \mathscr{B}, \mu)$ and $(X', \mathscr{B}', \mu')$. Thus, by Theorem 2.8, there is a complete, separating sequence $\{B'_i : i = 1, 2, \ldots\}$ of sets $B'_i \in \mathscr{B}'$ which is generating for $\mathscr{B}'$. Let

$$\xi_n := \{B'_1, \ldots, B'_n\} \cup \{X \setminus \bigcup_{i=1}^{n} B'_i\}, \quad n \geq 1.$$

It is clear that $\xi_1 < \xi_2 < \ldots$ and $\bigcup_{n \geq 1} \mathscr{B}(\xi_n)$ generates $\mathscr{B}'$. Now, define the countable partition

$$\xi' := \bigvee_{n=1}^{\infty} \xi_n$$

and set $\xi = \xi' \cup Y$, that is, ξ is the partition of X whose elements are elements of ξ' or atoms of μ. Since every element of ξ' is an element of ξ and since $\xi_n < \xi'$ for every n, we have that $\mathscr{B}' \subset \mathscr{B}(\xi)$. Also, by the definition of ξ, any atom of μ in $\mathscr{B}$ is an element of ξ, thus $Y \subset \mathscr{B}(\xi)$. Thus we have $\mathscr{B} = \mathscr{B}' \cup Y \subset \mathscr{B}(\xi) \subset \mathscr{B}$ which implies $\mathscr{B} = \mathscr{B}(\xi)$ as we wanted.

The construction just described establishes a way to define, from a given sigma algebra $\mathscr{B}$, a partition ξ such that $\mathscr{B}(\xi) = \mathscr{B}$. We will denote such partition ξ by $H(\mathscr{B})$.

Definition 2.11 Let $(X, \mathscr{A}, \mu)$ be a Lebesgue space and ξ be a partition of X. The partition $H(\mathscr{B}(\xi))$ is called the measurable Hull of ξ and is denoted by $\mathscr{H}(\xi)$.

It is an exercise to verify that measurability of a certain partition is equivalent to have an equivalence modulo zero of such partition with its measurable hull.

Exercise 2.1 Prove that a partition ξ of a Lebesgue space X is a measurable partition (see Definition 2.6) if, and only if, $\mathcal{H}(\xi)$ is equivalent modulo zero to ξ, that is, given any element $A \in \xi$ there exists an element $B \in \mathcal{H}(\xi)$ for which $\mu(B \Delta A) = 0$.

Let $T : X \to X$ be an automorphism of X.

Definition 2.12

(1) We say that a partition α of X is T-invariant if $T^{-1}\alpha < \alpha$. If $T^{-1}\alpha = \alpha$, we say that α is completely invariant.
(2) A partition ξ is exhaustive under an automorphism T if it is T-invariant and $\bigvee_{n=0}^{\infty} T^n \xi = \varepsilon$.

The following is a classical fact from entropy theory.

Theorem 2.12 (11.2 in [10]) *The following statements are equivalent for an automorphism:*

(a) the entropy is zero;
(b) the only exhaustive partition is the trivial partition by points ε.

Let $T : X \to X$ an automorphism of $(X, \mathcal{A}, \mu)$ and ξ be a partition of X with finite entropy. We denote by ξ^- the partition generated by the arbitrary future with respect to ξ, that is,

$$\xi^- := \bigvee_{n=0}^{\infty} T^{-n}\xi.$$

It is clear that $T^{-1}\xi^- < \xi^-$. Thus, if $h_\mu(T) = 0$, then ξ^- is completely invariant (see Theorem 11.1 from [10]). On the other hand, one can also prove (using the fact that $h_\mu(T, \xi) = \lim_n H_\mu(\xi | \bigvee_{j=1}^{n} T^{-j}\xi)$, see Proposition 2.5) that if $T^{-1}\xi^- = \xi^-$ then $h_\mu(T, \xi) = 0$. That is, T has zero entropy if, and only if, $T^{-1}\xi^- = \xi^-$. Denote

$$\Pi(\xi) := \mathcal{H}\left(\bigwedge_{n=1}^{\infty} T^{-n}\xi^-\right).$$

Observe that $\Pi(\xi) < \xi^-$.

Theorem 2.13 ([10]) *Let $T : X \to X$ an automorphism of a Lebesgue space $(X, \mathcal{A}, \mu)$ and ξ be a partition of X with finite entropy. If η is a partition of X satisfying $\eta < \Pi(\xi)$, then $h_\mu(T, \eta) = 0$.*

Definition 2.13 Let (X, μ) be a Lebesgue space and $T : (X, \mu) \to (X, \mu)$ a measure preserving automorphism. The family Π defined by

$$\Pi = \{A \subset X : h_\mu(T, \alpha) = 0, \alpha := \{A, X - A\}\}$$

is a σ-algebra called the Pinsker σ-algebra of T.

From the definition of Pinsker σ-algebra and from Theorem 2.13 we have that for every countable partition η, every element of $\Pi(\eta)$ belongs to Π. Thus,

$$\Pi(\eta) < \mathscr{H}(\Pi).$$

Theorem 2.14 (Theorem 11.5 in [10]) *For every automorphism $T : X \to X$ there exists a completely invariant partition $\pi = \pi(T)$, given by*

$$\pi = \mathscr{H}(\Pi)$$

where Π is the Pinsker σ-algebra of T, such that if η is any countable partition of X with finite entropy then

$$\eta < \pi \Leftrightarrow h_\mu(T, \eta) = 0.$$

The partition $\pi = \pi(T)$ given by the previous theorem is called the Pinsker partition of T.

Theorem 2.15 (Theorem 12.1 in [10]) *If T is an automorphism and ξ is an exhausting partition, then*

$$\bigwedge_0^\infty T^{-n} \xi > \pi(T).$$

Definition 2.14 Let (X, μ) be a measure space. We say that a measurable measure preserving map $f : X \to X$ has completely positive entropy if

$$h_\mu(T, \xi) > 0,$$

for any non-trivial finite partition ξ of X.

2.7 The Bernoulli Property

2.7.1 Bernoulli Shifts

Let us now see the Bernoulli Shifts. These are the most chaotic systems, from the ergodic point of view. In 1898 Jacques Hadamard analyzed the geodesics on some surfaces using infinity sequences, and that was the birth of Symbolic Dynamics.

Fixed a $d \in \mathbb{N}$ define the following set.

$$\Sigma_d := \{1, \ldots, d\}^{\mathbb{Z}} = \{(x_n)_{n \in \mathbb{Z}} \mid x_n \in \{1, \ldots, d\}\}$$

Given $a_n, \ldots, a_{n+m} \in \{1, \ldots, d\}$ we call by *cylinders* the sets defined by:

$$[n; a_n, \ldots, a_{n+m}] := \{(x_n)_{n\mathbb{Z}} \mid x_n = a_n, \; x_{n+1} = a_{n+1}, \ldots, x_{n+m} = a_{n+m}\}$$

Consider on Σ_d the σ-álgebra $\mathscr{B}_d$ generated by the family of all cylinders. Define a *Bernoulli measure* μ associated to $p_1, \ldots, p_d$ where $p_1 + \cdots + p_d = 1$, such that

$$\mu([n; a_n, \ldots, a_{n+m}]) = p_{a_n} \cdots p_{a_{n+m}}$$

or equivalently $\mu = \nu^{\mathbb{Z}}$, that is the product measure where ν is a measure on $\{1, \ldots, d\}$ defined by $\nu(\{i\}) = p_i$.

Now define the shift map

$$\sigma : \sum \rightarrow \sum$$
$$(x_n)_{n \in \mathbb{Z}} \mapsto (x_{n+1})_{n \in \mathbb{Z}}.$$

If you are getting into contact with Bernoulli shifts for the first time, you should think it as the toss of coins. Hence the measure defined above the natural measure to analyze a dice of d sides. We now have:

Theorem 2.16 *In the above notation, σ preserves the measure μ.*

Proof σ preserves the measure on cylinder since

$$\sigma^1([n; a_n, \ldots, a_{n+m}]) = [n + 1; a_n, \ldots, a_{n+m}].$$

The cylinders form an algebra and the set

$$\mathscr{A} = \{A \in \mathscr{B}_k \mid \mu(\sigma^{-1}(A)) = \mu(A)\}$$

is a monotone class which contains them. Therefore by the Monotone Class Theorem (see Theorem 3.4 of [1]) σ preserves μ. $\square$

What one should bear in mind is that usually to check some invariant condition with respect to a measure one should only look at the generators of the sigma algebra. That is what we do once again in the next result.

Theorem 2.17 $\forall A, B \in \mathscr{B}_d, \; \lim\limits_{n \to \infty} \mu(A \cap \sigma^{-m}(B)) = \mu(A)\mu(B).$

Proof We only need to prove for A, B cylinders. Let $A = [n; a_n, \ldots, a_{n+m}]$ and $B = [p; b_p, \ldots, a_{p+q}]$. Then $\sigma^{-m}(B) = [p + m; b_p, \ldots, b_{p+q}]$. Let m be such that $p + m + q \geq n + r$. Then we have the following disjoint union

$$A \cap \sigma^{-m}(B) = \bigcup_{c_1,\ldots,c_s \in \{1,\ldots,d\}} [n; a_n, \ldots, a_{n+m}, c_1, \ldots, c_s, b_p, \ldots, a_{p+q}],$$

where $s = p + m - n - r - 1$. Let us assume, to simplify notation, that $s = 2$. Thus

$$A \cap \sigma^{-m}(B) = \bigcup_{i=1}^{d} \bigcup_{j=1}^{d} [n; a_n, \ldots, a_{n+m}, c_i, c_j, b_p, \ldots, a_{p+q}].$$

Consequently

$$\mu(A \cap \sigma^{-m}(B)) = \sum_{i=1}^{d} \sum_{j=1}^{d} \mu([n; a_n, \ldots, a_{n+m}, c_i, c_j, b_p, \ldots, a_{p+q}])$$

$$= \mu(A)\mu(B) \sum_i p_i \sum_j p_j = \mu(A)\mu(B).$$

$\square$

Corollary 2.2 (σ, μ) *is mixing, therefore ergodic.*

Definition 2.15 We say that a dynamical system $f : X \to X$ preserving a measure ν is Bernoulli, or that it has the Bernoulli property, if there is a bijective measure preserving map $h : (X, \nu) \to (\Sigma, \mu)$ (where μ is the Bernoulli measure) and a set $A \subset X$ of full ν measure such that

$$\sigma \circ h(x) = h \circ f(x), \quad \forall x \in A.$$

In other words, f is said to be Bernoulli if it is measurably isomorphic (mod 0) (see Sect. 2.5) to a Bernoulli shift.

One of the most celebrated theorems involving classification of certain dynamical systems through dynamical invariants is the following theorem due to Ornstein in which he proves that Bernoulli shifts are completely classified by their metric entropy. The same is not true for Kolmogorov automorphisms.

Theorem 2.18 (Ornstein [5, 6]) *Two Bernoulli shifts* (σ_1, μ_1), (σ_2, μ_2) *are measurably isomorphic if, and only if, they have the same metric entropy. Consequently, given two probability spaces* (X, μ) *and* (Y, ν), *if* $f : X \to X$ *and* $g : Y \to Y$ *are measure preserving functions with the Bernoulli property, then* f *is measurably isomorphic to* g *if, and only if,*

$$h_\mu(f) = h_\nu(g).$$

2.7.2 Bernoulli Partitions

In this section we will see how one can characterize the Bernoulli property in terms of partitions.

Analogous to the case of Bernoulli shift we also define the distribution, or probability vector, of a finite partition α of X and the concept of independence between two partitions.

Given a finite partition $\alpha = \{A_1, \ldots, A_k\}$ the *distribution* or *probability vector* of α is the vector

$$d(\alpha) := (\mu(A_1), \ldots, \mu(A_k)).$$

Given a measurable automorphism $f : X \to X$ and $\alpha = \{A_1, \ldots, A_k\}$ a finite partition of X we denote by $f^{-1}(\alpha)$ the partition of X whose atoms are inverse images of the atoms of α and with the induced ordering, that is:

$$f^{-1}(\alpha) = \{f^{-1}(A_1), \ldots, f^{-1}(A_k)\}.$$

In case f is an isomorphism then we will denote by $f(\alpha)$ the partition given by the images, under f, of elements of α and with ordering induced by the ordering of α, that is,

$$f(\alpha) := \{f(A_1), \ldots, f(A_k)\}.$$

Definition 2.16 Two partitions α and β of X are independent if

$$\mu(A \cap B) = \mu(A) \cdot \mu(B), \quad A \in \alpha, \quad B \in \beta.$$

A sequence of partitions $\{\alpha_n\}_{n \geq 1}$ is said to be independent if for every $n > 1$ the partitions α_n and $\bigvee_1^{n-1} \alpha_i$ are independent.

The following proposition is an easy consequence of the uniqueness theorem on the extension of measures (see [3, Theorem A, p. 54]) and will be useful in the proof of the Theorem 2.19.

Proposition 2.6 Let $\phi : X_1 \to X_2$, where $(X_1, \mathscr{A}_1, \mu_1)$ and $(X_2, \mathscr{A}_2, \mu_2)$ are probability spaces. Let $\mathscr{E}$ be a generator for $\mathscr{A}_2$, and suppose

(i) $\phi^{-1}(\mathscr{E}) \subset \mathscr{A}_1$;
(ii) $\mu_1(\phi^{-1}(A)) = \mu_2(A)$, $A \in \mathscr{E}$.

Then $\phi^{-1}(\mathscr{A}_2) \subset \mathscr{A}_1$ and (ii) holds for all $A \in \mathscr{A}_2$.

The next theorem provides a very useful characterization of the Bernoulli property in terms of partitions.

Theorem 2.19 *Let* $(X, \mathscr{A}, \mu)$ *be a Lebesgue space. A measurable isomorphism* $f : X \to X$ *is isomorphic to the Bernoulli shift on k-symbols* $\sigma : \Sigma_k \to \Sigma_k$ *with distribution* $d(\sigma)$ *if and only if there exists a finite partition* $\alpha = \{A_1, \ldots, A_k\}$ *of* X *such that*

(1) $d(\alpha) = d(\sigma)$;
(2) α *is a generating partition for* f;
(3) $\{f^n \alpha\}_{n \geq 1}$ *is an independent sequence of partitions.*

Proof Let σ be the Bernoulli shift on k-symbols and with distribution $d(\sigma) = (p_1, \ldots, p_k)$. Let $\mathscr{C}$ be the partition given by the two cylinders $C_0 = \{(x_n)_{n \in \mathbb{N}} : x_0 = 0\}$ and $C_1 = \{(x_n)_{n \in \mathbb{N}} : x_0 = 1\}$. We have already showed in the proof of Theorem 2.21 that $\mathscr{C}$ is a generating partition for σ and it is clear that the sequence of partitions $\{f^n(\mathscr{C})\}_n$ is independent since every atom of such partitions are cylinders. Thus $\mathscr{C}$ is a partition of Σ_k satisfying (1)–(3). Now, if f is isomorphic to the Bernoulli shift by a measure isomorphism $h : X \to \Sigma_k$, then the partition $\alpha := h^{-1}(\mathscr{C})$ will satisfy (1)–(3) for f. This proves one side of the statement.

To prove the converse, let $\alpha = \{A_1, A_2, \ldots, A_k\}$ be a partition satisfying (1)–(3) for a map f. Define $h : X \to \Sigma_k$ by

$$h(x) = (x_n)_n, \quad \text{such that } f^n(x) \in A_{x_n}.$$

It is easy to see that if $h(x) = h(y)$ then, for each $n \in \mathbb{Z}$, the points $f^n(x)$ and $f^n(y)$ belong to the same element of α. Also, it is clear that $h \circ f(x) = \sigma \circ h(x)$, therefore we just need to prove that h is a measure preserving isomorphism *mod* 0.

Claim 1 h is injective almost everywhere. Since α is a generating partition for f, the countable collection of sets $\bigcup_{-\infty}^{\infty} f^n(\alpha)$ generates $\mathscr{A}$ and then, by Theorem 2.9, it also separates X. In other words there exists a set $E \subset X$ with $\mu(E) = 0$ such that, given any two distinct points $x, y \in X \setminus E$ there exists a set $S \in \bigcup_{-\infty}^{\infty} f^n(\alpha)$ such that S contains exactly one of the points x, y, in particular, for some $n \in \mathbb{Z}$ the points $f^n(x)$ and $f^n(y)$ are in distinct points of α which implies that $h(x) \neq h(y)$. Therefore, h is injective on $X \setminus E$.

Claim 2 h is measure preserving. By Proposition 2.6 it is enough to prove that h preserves the measures of all of the sets of the form $h^{-1}(C)$, C cylinder in Σ. Take a general cylinder $C = \{x \in \Sigma_k : x_i = t_i, -m \leq i \leq n\}$. By the definition of h we have

$$h^{-1}(C) \overset{\circ}{=} \bigcap_{i=-m}^{n} f^i(A_{t_i}).$$

Now by item (3) we have

$$\mu(h^{-1}(C)) = \prod_{-m}^{n} \mu(f^i(A_{t_i})) = \prod_{i=-m}^{n} \mu_\sigma(\{x \in \Sigma_k : x_i = t_i\}) = \mu_\sigma(C).$$

Thus, h is a measurable and measure preserving. Finally, since the image of h must have full measure, it follows from Theorem 2.10 that h is a measurable isomorphism mod 0 as we wanted to show. $\square$

In the light of Theorem 2.19 we will call a finite partition α a Bernoulli partition if it satisfies the items (2) and (3) of the theorem.

Definition 2.17 Let $(X, \mathscr{A}, \mu)$ be a Lebesgue space and $f : X \to X$ a μ-invariant map. We say that a partition α of X is a Bernoulli partition if α is generating and independent with respect to f, i.e., $\{f^n \alpha\}_{n \geq 1}$ is an independent sequence of partitions.

2.8 The Kolmogorov Property

In this section we introduce the Kolmogorov property. This is a more technical definition and that is why we left it for the last section. Among the hierarchy of ergodic properties the Kolmogorov property is weaker than the Bernoulli property (i.e., being a Bernoulli systems). Theorem 2.21 proves that a Bernoulli system is Kolmogorov and a great deal of effort is being done to understand when the Kolmogorov and Bernoulli properties coincide.

Definition 2.18 Given a partition α of a probability space (X, μ), we say that a property P holds for ε-almost every atom $A \in \alpha$ if the union of all atoms of α for which P does not hold has measure less than ε.

Definition 2.19 We say that a finite partition ξ of X is a Kolmogorov partition for a measure preserving automorphism $T : X \to X$ if given any $B \in \bigvee_{-\infty}^{+\infty} T^{-k}\xi$ and $\varepsilon > 0$ there exists an $N_0 = N(\varepsilon, B)$ such that for all $N' \geq N \geq N_0$ and ε-almost every atom A of $\bigvee_N^{N'} T^k \xi$ we have

$$\left| \frac{\mu(A \cap B)}{\mu(A)} - \mu(B) \right| \leq \varepsilon.$$

The following very strong Theorem, due to Rohlin and Sinai [9], shows that Kolmogorov property is an enough and sufficient condition to have uncertainty of a system in the sense that every non-trivial finite partition generates a positive information, that is, we obtain a relation between the Kolmogorov property of T and the Pinsker σ-algebra of T.

Theorem 2.20 (Rohlin–Sinai [9]) *Let $(X, \mathscr{A}, \mu)$ be a Lebesgue space and $T : (X, \mu) \to (X, \mu)$ be a measure preserving automorphism. Then, there exists a σ-algebra $\mathscr{K} \subset \mathscr{A}$ with*

(i) $T^{-1}\mathscr{K} \subset \mathscr{K}$;
(ii) $\bigvee_{n=0}^{\infty} T^{-n}\mathscr{K} = \mathscr{A}$;

(iii) $\bigcap_{n=0}^{\infty} T^n \mathscr{K} = \Pi$, *where Π denotes the Pinsker σ-algebra of T.*

In particular,

$$\Pi = \bigcap_{m=1}^{\infty} \bigvee_{j=m}^{\infty} T^j \alpha,$$

where α is a generating partition, and T is a K-system if, and only if, its Pinsker σ-algebra is trivial, that is, T has completely positive entropy.

Corollary 2.3 *T is a K-system if, and only if, its Pinsker partition $\pi(T)$ is trivial.*

The next lemma will be useful to prove that Kolmogorov automorphisms of $\mathbb{T}^2$ are Bernoulli.

Lemma 2.1 ([7, 9]) *Let T be a K-automorphism of a Lebesgue space (Y, μ), then every finite partition ξ of Y is a Kolmogorov partition.*

Let us now position the Kolmogorov property in the hierarchy of the ergodic properties by proving that the K-property is stronger than mixing but weaker than Bernoulli property.

Theorem 2.21 *If $T : X \to X$ is a Bernoulli automorphism in a Lebesgue space $(X, \mathscr{A}, \mu)$, then T is a K-automorphism.*

Proof It is enough to prove that Bernoulli shifts are K-automorphisms. Let $\sigma : \Sigma \to \Sigma$ be a Bernoulli shift where $\Sigma = \{1, \ldots, k\}^{\mathbb{Z}}$ is endowed with a Bernoulli measure μ, induced by a probability vector $p = (p_1, \ldots, p_k)$, and the σ-algebra $\mathscr{B}$ generated by the cylinders of Σ. Consider the partition ξ given by $\xi = \{[0 : a] : 1 \le a \le k\}$. Let us prove that ξ is a Kolmogorov partition for σ. Observe that $\sigma^{-n}\xi = \{[n : a] : 1 \le a \le k\}$, thus

$$\bigvee_{i=-n}^{m} \sigma^{-i}\xi = \{[-n : a_0, a_1, \ldots, a_{n+m}] : 1 \le a_i \le k, \text{ for } 0 \le i \le n+m\}.$$

Consequently, any σ-algebra containing $\bigcup_{n=-\infty}^{\infty} \sigma^n \alpha$ must contain all the cylinders of Σ, which implies that it contains $\mathscr{A}$. Thus, ξ is a generating partition. Now, let $B \in \mathscr{A}$ an arbitrary measurable set. Given any $N' > N$ let $A \in \bigvee_{N}^{N'} \sigma^i \xi$ be any atom of positive measure. Given any $\varepsilon > 0$ there exists a cylinder C such that

$$\mu(B \triangle C) < \frac{1}{2}\varepsilon \cdot \mu(A).$$

Since A is also a cylinder we have $\mu(A \cap C) = \mu(A) \cdot \mu(C)$. Thus

$$\left| \frac{\mu(A \cap B)}{\mu(A)} - \mu(B) \right| = \left| \frac{\mu(A \cap C)}{\mu(A)} - \frac{\mu(A \cap (C \setminus B))}{\mu(A)} \right.$$
$$\left. + \frac{\mu(A \cap (B \setminus C))}{\mu(A)} - \mu(B) \right|$$
$$\leq \left| \frac{\mu(A \cap C)}{\mu(A)} - \mu(B) \right| + \frac{\mu(C \setminus B) + \mu(B \setminus C)}{\mu(A)}$$
$$< |\mu(C) - \mu(B)| + \frac{1}{2}\varepsilon < \varepsilon.$$

Thus, ξ is indeed a Kolmogorov partition for σ and σ is a K-automorphism as we wanted to show. $\qquad\square$

Theorem 2.22 *If $T : X \rightarrow X$ is a K-automorphism in a Lebesgue space* $(X, \mathscr{A}, \mu)$, *then T is mixing.*

Proof The proof we present here is adapted from the proof of Theorem 3.53 in [3] where he actually proves a more general fact, namely that the K-property implies the uniformly mixing property.

By Theorem 2.20, if T is a K-automorphism then its Pinsker σ-algebra is trivial and, consequently, there exists a σ-algebra $\mathscr{K}$ for which:

(i) $T^{-1}\mathscr{K} \subset \mathscr{K}$;
(ii) $\bigvee_{n=0}^{\infty} T^{-n}\mathscr{K} = \mathscr{A}$;
(iii) $\bigcap_{n=0}^{\infty} T^{n}\mathscr{K}$ is the trivial σ-algebra.

By (ii) it is enough to prove that

$$\lim_{n \to \infty} \mu(A \cap T^{n}(B)) = \mu(A)\mu(B) \qquad (2.2)$$

for every $A \in \mathscr{A}$ and $B \in \bigvee_{i=1}^{\infty} T^{-i}\mathscr{K}$. Assume by contradiction that there exist $A \in \mathscr{A}$ and $B \in \bigvee_{i=1}^{\infty} T^{-i}\mathscr{K}$ for which (2.2) does not hold. Consider the function $\rho : \mathscr{A} \rightarrow \mathbb{R}$ given by

$$\rho(C) := \mu(A \cap C) - \mu(A)\mu(C).$$

ρ is a finite signed measure. Let $\mathscr{A}_n$ be the smallest sigma-algebra containing $\bigcup_{j=n}^{\infty} T^{j}B$ and define $\rho_n = \rho|\mathscr{A}_n$. Consider C_n to be the maximal positive set of ρ_n, that is, $\rho_n(E) \geq 0$ for every $E \in \mathscr{A}_n$ with $E \subset C_n$. Since (2.2) is not satisfied there is a $\delta > 0$ for which

$$\rho_n(C_n) \geq \delta, \quad \forall n \geq 1.$$

Now, for $m \geq n$ we have $C_m \in \mathscr{A}_n$ and

$$\rho_n(C_m \cup C_{m-1} \cup \ldots \cup C_n)$$

$$= \rho_n(C_m) + \rho_n(C_{m-1} \setminus C_m) + \rho_n(C_{m-2} \setminus (C_{m-1} \cup C_m))$$

$$+ \ldots + \rho_n \left(C_n \setminus \bigcup_{j=n+1}^{m} C_j \right)$$

$$= \rho_m(C_m) + \rho_{m-1}(C_{m-1} \setminus C_m) + \ldots + \rho_n \left(C_n \setminus \bigcup_{j=n+1}^{m} C_j \right) \geq \rho_m(C_m) \geq \delta.$$

Denote $D_n = \bigcup_{m=n}^{\infty} C_m$ and $D = \bigcap_n D_n$. Thus we have $\rho_n(D_n) \geq \delta$ for every $n \geq 1$ and

$$\mu(A \cap D) - \mu(A)\mu(D) = \lim_{n \to \infty} (\mu(A \cap D_n) - \mu(A)\mu(D_n)) = \lim_{n \to \infty} \rho_n(D_n) \geq \delta.$$

In particular $0 < \mu(D) < 1$. But observe that since $C_n \in \mathscr{A}_n$ for every n then $D \in \bigcap_{n=1}^{\infty} T^{n-m} \mathscr{K}$ for some $m \geq 1$. By property (iii) of the σ-algebra $\mathscr{K}$ we have $\bigcap_{n=1}^{\infty} T^{n-m} \mathscr{K}$ is the trivial σ-algebra, from where we obtain a contradiction with the fact that D does not have either zero or full μ-measure. Therefore T is indeed mixing as we wanted.   □

References

1. Billingsley, P.: Probability and Measure. Wiley Series in Probability and Mathematical Statistics, 3rd edn. Wiley, New York (1995)
2. Einsiedler, M.,Ward, T.: Ergodic Theory. Springer, London (2013)
3. Glasner, E.: Ergodic Theory via Joinings. Mathematical Surveys and Monographs, vol. 101. American Mathematical Society, Providence (2003)
4. Kechris, A.S.: Classical Descriptive Set Theory. Graduate Texts in Mathematics, vol. 156. Springer, New York (1995)
5. Ornstein, D.S.: Two Bernoulli shifts with infinite entropy are isomorphic. Adv. Math. **5**, 339–348 (1970)
6. Ornstein, D.S.: Imbedding Bernoulli shifts in flows. In: Contributions to Ergodic Theory and Probability, Lecture Notes in Mathematics, vol. 160, pp. 178–218. Springer, Berlin (1970)
7. Pinsker, M.S.: Dynamical systems with completely positive or zero entropy. Soviet Math. Dokl. **1**, 937–938 (1960)
8. Rohlin, V.A.: On the fundamental ideas of measure theory. Am. Math. Soc. Transl. **71**, 1–54 (1952)
9. Rohlin, V.A., Sinaĭ, J.G.: Construction and properties of invariant measurable partitions. Dokl. Acad. Nauk. SSSR. **141**, 1038–1041 (1961)
10. Rokhlin, V.A.: Lectures on the entropy theory of measure-preserving transformations. Russ. Math. Surv. **22**(5), 1–52 (1967)
11. Walters, P.: An Introduction to Ergodic Theory, vol. 79. Springer Science & Business Media, New York (2000)

Chapter 3
Kolmogorov–Bernoulli Equivalence for Ergodic Automorphisms of $\mathbb{T}^2$

Abstract The main goal of this chapter is to show that the class of ergodic automorphisms of the 2-torus are Bernoulli. The proof summarized in this chapter was originally given by Ornstein and Weiss in 1973 in the article entitled "Geodesic flows are Bernoullian" (Ornstein and Weiss, Isr J Math 14:184–198, 1973). The method introduced by Ornstein–Weiss uses the geometric structures associated to the ergodic automorphisms of $\mathbb{T}^2$ to obtain a sequence of refining partitions which are Very Weak Bernoulli (VWB) so that, by Ornstein Theory, one concludes that the automorphism is actually Bernoulli. The same method is used in Ornstein and Weiss (Isr J Math 14:184–198, 1973) to show that geodesic flows on negatively curved Riemannian surfaces are Bernoulli. Posteriorly many authors used the tools introduced by Ornstein and Weiss to show that the Kolmogorov property implies Bernoulli property for a larger class of dynamics such as the Anosov diffeomorphisms which will be treated later in this book. Until now, the ideas introduced by Ornstein and Weiss are still in the core of the arguments used to obtain Bernoulli property from the Kolmogorov property.

In Sect. 3.1 we introduce the concept of very weak Bernoulli partitions and state two theorems of Ornstein theory (Theorems 3.2 and 3.3) which are crucial in the proof of the main theorems of this chapter and Chap. 4. In Sect. 3.2 we show that ergodic automorphisms of $\mathbb{T}^2$ are Kolmogorov by referring to a more general result proven in Chap. 3. Finally, in Sect. 3.3 we show in detail the main result of this chapter, namely we show that ergodic automorphisms of $\mathbb{T}^2$ are Bernoulli. This section may be considered as the most important section of this book since the argument showed in this section is essentially the same argument used multiple times in the theory to show that Kolmogorov diffeomorphisms with certain hyperbolic structure are Bernoulli.

3.1 Finite and Very Weak Bernoulli Partitions

In the seventies Ornstein obtained a remarkable result when he proved that Bernoulli shifts are completely classified by their entropy. To do so, Ornstein introduced the concept of Very Weak Bernoulli partitions which is in the core of Ornstein's Theory. Roughly speaking the concept of Very Weak Bernoulli partitions allows us to verify the Bernoulli property for a certain system without properly finding a conjugacy of the given system with a Bernoulli shift. Instead, Ornstein's results show that it is enough to construct partitions which are arbitrarily close to be a Bernoulli partition.

3.1.1 The $\bar{d}$-Distance in the Space of Finite Partitions

To define a concept of distance in the space of finite partitions of certain Lebesgue spaces we will consider all the measures on the product spaces whose projections are the original measures, and then we measure how big is the set of pairs which belong to atoms of different indexes. To put this idea in formal terms we will make use of the concept of joining of two measure spaces. Though there is a beautiful and deep theory of joinings of measure spaces, in this book we will only use the basic definition of joining. We refer the interested reader to [2] for an exposition on joinings and ergodic theory.

Definition 3.1 Let $\mathbf{X} = (X, \mu)$ and $\mathbf{Y} = (Y, \nu)$ be non-atomic Lebesgue probability spaces. We say that a probability measure λ on the product space $X \times Y$ is a joining of X and Y if the marginals, or projections, of λ are μ and ν, that is, for any measurable sets $A \subset X$, $B \subset Y$ we have

$$\lambda(A \times Y) = \mu(A), \quad \text{and} \quad \lambda(X \times B) = \nu(B).$$

We denote by $J(\mathbf{X}, \mathbf{Y})$ the set of all joinings of $\mathbf{X} = (X, \mu)$ and $\mathbf{Y} = (Y, \nu)$.

Let $\alpha = \{A_1, \ldots, A_k\}$ and $\beta = \{B_1, \ldots, B_k\}$ be finite partitions of the non-atomic Lebesgue probability spaces $\mathbf{X} = (X, \mu)$ and $\mathbf{Y} = (Y, \nu)$ respectively. Given $x \in X$, denote by $\alpha(x)$ the atom of α which contains x. For $y \in Y$, $\beta(y)$ is defined in a similar way.

Definition 3.2 The $\bar{d}$-distance between α and β is defined by:

$$\bar{d}(\alpha, \beta) = \inf_{\lambda \in J(\mathbf{X}, \mathbf{Y})} \lambda\{(x, y) : \alpha(x) \neq \beta(y)\}.$$

Observe that the definition of the $\bar{d}$-distance reflects the idea given in the first paragraph of this section, that is, we consider a measure $\lambda \in J(\mathbf{X}, \mathbf{Y})$ (which is a measure in $X \times Y$) and we take the λ-measure of the set of pairs belonging to atoms of different indexes. Then we take the infimum over $\lambda \in J(\mathbf{X}, \mathbf{Y})$. The following simple examples provide a visualization of this concept.

Fig. 3.1 The gray set is the set of pairs whose coordinates belong to atoms of same index

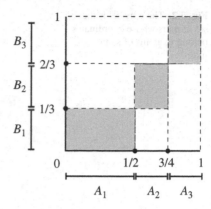

Example 3.1 Let $X = Y = I = [0, 1]$ endowed with the standard Lebesgue measure m. Take the partitions

$$\alpha = \{[0, 1/2), [1/2, 3/4), [3/4, 1]\} \quad \text{and} \quad \beta = \{[0, 1/3), [1/3, 2/3), [2/3, 1]\}.$$

Let us find $\overline{d}(\alpha, \beta)$. Consider the space $I \times I$ as in Fig. 3.1 above.

The gray set $D = (A_1 \times B_1) \cup (A_2 \times B_2) \cup (A_3 \times B_3)$ depicted in the figure is the diagonal set, that is, it is the set of pairs $(x, y) \in I \times I$ for which $\alpha(x) = \beta(y)$. To evaluate $\overline{d}(\alpha, \beta)$ we need to estimate $\lambda(I \times I - D) = 1 - \lambda(D)$ for $\lambda \in J(\mathbf{I}, \mathbf{I})$. Let λ_0 be the probability measure given by:

- if $B \subset A_1 \times B_1 = [0, 1/2) \times [0, 1/3)$, then $\lambda_0(B) = \frac{1}{m(B_1)}(m \times m)(B)$;
- if $B \subset A_2 \times B_2 = [1/2, 3/4) \times [1/3, 2/3)$, then $\lambda_0(B) = \frac{1}{m(B_2)}(m \times m)(B)$;
- if $B \subset A_3 \times B_3 = [3/4, 1] \times [2/3, 1]$, then $\lambda_0(B) = \frac{1}{m(B_3)}(m \times m)(B)$.
- for any $B \subset I \times I$, $\lambda_0(B) := \sum_{j=1}^{3} \lambda_0(B \cap (A_j \times B_j))$.

That is, λ_0 is the measure in $I \times I$ such that restricted to $A_j \times B_j$ it is the product measure normalized to given measure $m(A_j)$ to $A_j \times B_j$, $j = 1, 2, 3$. It is easy to see that the projections of λ_0 in both coordinates are m and, by the definition of λ_0, we have

$$\lambda_0(I \times I - D) = 1 - \lambda_0(D) = 1 - \sum_{i=1}^{3} \lambda_0(A_j \times B_j) = 1 - \sum_{i=1}^{3} m(A_j) = 0.$$

Thus $\lambda_0 \in J(\mathbf{I}, \mathbf{I})$ and

$$\overline{d}(\alpha, \beta) = \inf_{\lambda \in J(\mathbf{I}, \mathbf{I})} \lambda\{I \times I - D\} \leq \lambda_0(I \times I - D) = 0 \Rightarrow \overline{d}(\alpha, \beta) = 0.$$

It is important for the reader to note that the infimum is not attained at the trivial product measure $\lambda := m \times m \in J(\mathbf{I} \times \mathbf{I})$.

Fig. 3.2 The gray set is the set of pairs whose coordinates belong to atoms of same index

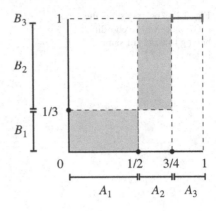

Example 3.2 Let $X = Y = I = [0, 1]$ endowed with the standard Lebesgue measure m. Let α be the same partition as in Example 3.1 and take β to be given by

$$\beta = \{[0, 1/3), [1/3, 1), \{1\}\}.$$

We repeat the same procedure. Let D be the diagonal painted in gray in Fig. 3.2.

Let $D = ([0, 1/2) \times [0, 1/3)) \cup ([1/2, 3/4) \times [1/3, 1)) \cup ([3/4, 1] \times \{1\})$. Observe that in this example we cannot repeat ipsis litteris the construction of λ_0 as done in Example 3.1 since $m \times m(A_3 \times B_3) = m \times m([3/4, 1] \times \{1\}) = 0$. Thus, in this case, we will use the product measure $\lambda := m \times m$ to get an estimative for $\overline{d}(\alpha, \beta)$. We have

$$m \times m(I \times I - D) = \frac{1}{2} \cdot \frac{2}{3} + \frac{1}{4} \cdot \frac{1}{3} + \frac{1}{4} \cdot 1 = \frac{2}{3}.$$

Thus

$$\overline{d}(\alpha, \beta) = \inf_{\lambda \in J(\mathbf{I}, \mathbf{I})} \lambda\{I \times I - D\} \leq m \times m(I \times I - D) = \frac{2}{3}.$$

The essential difference between Examples 3.1 and 3.2 is that in the first one, by taking the continuous piecewise linear map $\theta : [0, 1] \to [0, 1]$ mapping $[0, 1/2]$ to $[0, 1/3]$, $[1/2, 3/4]$ to $[1/3, 2/3]$ and $[3/4, 1]$ to $[2/3, 1]$, and taking $\lambda_1 = (\theta^{-1})_* m$, we see that the partitions α and β are actually equivalent by the isomorphism θ which is measure preserving between the spaces $([0, 1], \lambda_1)$ and $([0, 1], m)$, while in the latter one α and β are not equivalent in this sense. That is, in the first case by taking the measure λ_0 to be the probability measure in $I \times I$, supported in the set $\bigcup_{j=1}^{3} A_j \times B_j$ and normalized to give weight $m(A_j)$ to $A_j \times B_j$ we conclude that $\lambda_0 \in J(\mathbf{I}, \mathbf{I})$ and that $\overline{d}(\alpha, \beta) = 0$. In the second example, there is no such function since $m(A_3) > 0$ and $m(B_3) = 0$. With this idea in mind, one may define an equivalence relation on the set of finite partitions and then use this relation to define the $\overline{d}$-distance between partitions. Indeed, in [7] the authors define an equivalence class of partitions (resp. sequences of partitions) which have the same distribution, and use this relation to define the $\overline{d}$-distance between partitions (resp. sequences of partitions). We do not formalize this idea here since it will not be important along the rest of the presentation and refer the reader to [7] where this equivalent definition of the $\overline{d}$-distance is presented in a very concise way.

Definition 3.3 Given a sequence of finite partitions $\{\alpha_i\}_1^n$ of a non-atomic Lebesgue space (X, μ), we define the sequence of integer functions $l_i(x)$ by the condition $x \in A_{l_i(x)}^{(i)}$, where $\alpha_i = \{A_1^{(i)}, A_2^{(i)}, \ldots, A_{n_i}^{(i)}\}$. We call the sequence of functions $l_i(x)$ the α-name of the sequence of partitions $\{\alpha_i\}_1^n$.

Given two sequences of finite partitions $\{\alpha_i\}_{i=1}^n$ and $\{\beta_i\}_{i=1}^n$ of X and Y respectively, a natural way to measure the difference between the α-name of a point $x \in X$ and the β-name of a point $y \in Y$ is to take the function

$$h(x, y) = \frac{1}{n} \sum_{i:l_i(x) \neq m_i(y)} 1, \tag{3.1}$$

where $\{l_i\}_{i=1}^n$ is the α-name of the sequence of partitions $\{\alpha_i\}_1^n$ and $\{m_i\}_{i=1}^n$ is the β-name of the sequence of partitions $\{\beta_i\}_1^n$.

We can now extend the definition of the $\overline{d}$-distance to sequences of partitions.

Definition 3.4 The $\overline{d}$-distance between the sequences of finite partitions $\{\alpha_i\}_{i=1}^n$ and $\{\beta_i\}_{i=1}^n$ is defined by

$$\overline{d}(\{\alpha_i\}_{i=1}^n, \{\beta_i\}_{i=1}^n) = \inf_{\lambda \in J(\mathbf{X}, \mathbf{Y})} \int_{X \times Y} h(x, y) d\lambda.$$

Definition 3.5 We say that a transformation $\theta : X \to Y$ of Lebesgue spaces (X, μ) and (Y, ν) is ε-measure preserving if there exists a subset $E \subset X$ such that $\mu(E) \leq \varepsilon$ and for every measurable set $A \subset X \setminus E$,

$$\left| \frac{\nu(\theta(A))}{\mu(A)} - 1 \right| \leq \varepsilon. \tag{3.2}$$

Remark 3.1 In the literature (for example, in [1]) a ε-measure preserving transformation is sometimes defined with the inverse of the ratio considered in (3.2). However, it is easy to see that if θ is ε-measure preserving in the sense of [1] then it is $\varepsilon/(1 - \varepsilon)$-measure preserving with respect to the Definition 3.5. Thus, modulo changing some constants along the proof, it is indifferent to work with either one of these two definitions or the other.

The next lemma states that given a finite partition α of X, an ε-measure preserving function $\theta : X \to Y$ may be approximated by a measurable function which preserves the measure of atoms of α. The proofs of the next two lemmas are inspired in the proof of Lemma 4.3 in [1] in which they prove the same results under another, but equivalent, definition of ε-measure preserving functions (see Remark 3.1).

Lemma 3.1 *Let $\theta : X \to Y$ be an ε-measure preserving map between the Lebesgue. spaces (X, μ) and (Y, ν). Let $\alpha = \{A_1, \ldots, A_n\}$ be a finite partition of X. Then, there exists a map $\overline{\theta} : X \to Y$ such that the images $\overline{\theta}(A_j), j = 1, \ldots, n$ are mutually disjoint, $\mu(A_j) = \nu(\overline{\theta}(A_j))$ and*

$$\mu(\{x : \overline{\theta}(x) \neq \theta(x)\}) < C\varepsilon.$$

for a certain constant $C > 0$.

Proof Given a measurable set $S \subset X$ denote by $\overline{S}$ the set $S \backslash E$. Since θ is ε-measure preserving we have that

$$|\mu(\overline{S}) - \nu(\theta(\overline{S}))| < \varepsilon \mu(\overline{S}).$$

Thus,

$$\sum_j \nu(\theta(\overline{A}_j)) < (1 + \varepsilon) \sum_j \mu(\overline{A}_j) < 1 + 2\varepsilon \tag{3.3}$$

and

$$\nu \left(\bigcup_j \theta(\overline{A}_j) \right) \geq (1 - \varepsilon)\mu \left(\bigcup_j \overline{A}_j \right) = (1 - \varepsilon)\mu(X \setminus E) > 1 - 2\varepsilon. \tag{3.4}$$

Let

$$B := \theta^{-1} \left(\bigcup_{j \neq k} \theta(\overline{A}_j) \cap \theta(\overline{A}_k)) \right).$$

From (3.3) and (3.4) we have

$$\nu \left(\bigcup_{j \neq k} \theta(\overline{A}_j) \cap \theta(\overline{A}_k) \right) \leq \sum_j \nu(\theta(\overline{A}_j)) - \nu \left(\bigcup_j \theta(\overline{A}_j) \right) \leq 4\varepsilon.$$

Thus using again the fact that θ is ε-measure preserving we have

$$\mu(\overline{B})(1 - \varepsilon) < \nu(\theta(\overline{B})) \leq 4\varepsilon \Rightarrow \mu(\overline{B}) < \frac{4\varepsilon}{1 - \varepsilon} < c \cdot \varepsilon,$$

for a certain constant $c > 0$.

Now consider the sets $\widetilde{A}_j := A_j \setminus (B \cup E)$. Notice that $\theta(\widetilde{A}_j) \cap \theta(\widetilde{A}_k) = \emptyset$, for all $A_j \neq A_k$, and that

$$\mu\left(\bigcup_j \widetilde{A}_j\right) = \mu\left(\left[\bigcup \overline{A_j}\right] \setminus \overline{B}\right) = \mu(X \setminus E) - \mu(\overline{B}) > 1 - \varepsilon - c\varepsilon = 1 - (c+1)\varepsilon.$$

Let us now construct the map $\overline{\theta}$.

First Step   Consider those sets $\widetilde{A}_j$ where $\nu(\theta(\widetilde{A}_j)) > \mu(\widetilde{A}_j)$. Consider a set $G_j \subset \widetilde{A}_j$ which has the property that $\nu(\theta(G_j)) = \mu(\widetilde{A}_j)$. Notice that

$$\mu(G_j) > \frac{1}{1+\varepsilon}\nu(\theta(G_j)) = \frac{1}{1+\varepsilon}\mu(\widetilde{A}_j).$$

Define $\overline{\theta}$ to be equal to θ in G_j, and to map $\widetilde{A}_j \setminus G_j$ to any set of measure zero.

Second Step   Now consider those sets $\widetilde{A}_j$ where $\nu(\theta(\widetilde{A}_j)) < \mu(\widetilde{A}_j)$. Consider a set $G_j \subset \widetilde{A}_j$ whose measure is greater than $\frac{1}{1+\varepsilon}\mu(\widetilde{A}_j)$. Define $\overline{\theta}$ to be equal to θ in G_j, and to map $\widetilde{A}_j \setminus G_j$ to any set of measure $\mu(\widetilde{A}_j) - \nu(\theta(G_j))$ which does not intersect with the images of any of the other $\widetilde{A}_j$'s.

Clearly $\overline{\theta}$ satisfies the first property, that is, the images $\overline{\theta}(A_j)$ are mutually disjoint. Furthermore, $\{x : \overline{\theta}(x) \neq \theta(x)\} \subset X \setminus \bigcup_j G_j$ so

$$\mu(\{x : \overline{\theta}(x) \neq \theta(x)\}) < 1 - \frac{1}{1+\varepsilon}\mu\left(\bigcup_j \widetilde{A}_j\right) < \frac{2+c}{1+\varepsilon} \cdot \varepsilon < C\varepsilon,$$

for a certain constant $C > 0$, as we wanted to show. $\square$

The strategy to evaluate the $\bar{d}$-distance between two sequences of partitions is given by the following lemma which provides a way to estimate the $\bar{d}$-distance by comparing the names of points with respect to the given sequences of partitions. The proof given here is inspired in the proofs of Lemma 4.3 in [1] and Lemma 1.3 in [7].

Lemma 3.2   *Let (X, μ) and (Y, ν) be two nonatomic Lebesgue probability spaces. Let $\{\alpha_i\}$ and $\{\beta_i\}$, $1 \leq i \leq n$, two sequences of finite partitions of X and Y, respectively. Suppose there is a map $\theta : X \to Y$ such that*

(i) There is a set $E_1 \subset X$ with $\mu(E_1) < \varepsilon$, and

$$h(x, \theta(x)) < \varepsilon, \quad \forall x \in X \setminus E_1;$$

(ii) *There is a set* $E_2 \subset X$, *with* $\mu(E_2) < \varepsilon$, *such that for any measurable set* $A \subset X \setminus E_2$

$$\left| \frac{\nu(\theta(A))}{\mu(A)} - 1 \right| < \varepsilon.$$

Then

$$\bar{d}(\{\alpha_i\}, \{\beta_i\}) < c\varepsilon,$$

for a certain constant c independent of n and of the partitions.

Proof The idea is to construct a joining $\lambda \in J(X, Y)$ such that $\lambda(\{(x, y) : h(x, y) < \varepsilon\}) > 1 - D\varepsilon$ for a certain constant $D > 0$. Once constructed such a joining λ, by the definition of the $\bar{d}$-distance we will obtain

$$\bar{d}(\{\alpha_i\}_{i=1}^n, \{\beta_i\}_{i=1}^n) \leq \int_{X \times Y} h(x, y) d\lambda$$

$$= \int_{\{(x,y):h(x,y)<\varepsilon\}} h(x, y) d\lambda + \int_{X \times Y - \{(x,y):h(x,y)<\varepsilon\}} h(x, y) d\lambda$$

$$\leq \varepsilon + (1 - \lambda(\{(x, y) : h(x, y) < \varepsilon\})) < \varepsilon + D\varepsilon = (1 + D)\varepsilon,$$

and then, by taking $c := 1 + D$, the claim is proved.

Let $\alpha := \bigvee_{i=1}^n \alpha_i =: \{A_1, A_2, \ldots, A_k\}$ and apply Lemma 3.1 to the partition α. We obtain a function $\bar{\theta} : X \to Y$ such that $\bar{\theta}(A_j)$, $j = 1, \ldots, k$, are pairwise disjoint and

$$\mu(\{x \in X : \theta(x) \neq \bar{\theta}(x)\}) < C \cdot \varepsilon,$$

for a certain constant $C > 0$. We now construct the desired joining λ in a similar way to what we have done in Example 3.1.

Let λ be the measure on $X \times Y$ supported on the set $\bigcup_{j=1}^n A_j \times \bar{\theta}(A_j)$ and such for $B \subset A_j \times \bar{\theta}(A_j)$ the measure λ is defined by:

$$\lambda(B) = \frac{1}{\mu(A_j)} \mu \times \nu,$$

that is, restricted to $A_j \times \bar{\theta}(A_j)$, λ is given by the product measure normalized to give weight $\mu(A_j) = \nu(\bar{\theta}(A_j))$ to $A_j \times \bar{\theta}(A_j)$. In particular, as $\bar{\theta}(A_j)$ are disjoint we have that if $B \subset A_j$ then $\lambda(B \times Y) = \lambda(B \times \bar{\theta}(A_j)) = \mu(B)$, $j = 1, \ldots, k$, and similarly if $C \subset \bar{\theta}(A_j)$ then $\lambda(X \times C) = \lambda(A_j \times C) = \nu(C)$, $j = 1, \ldots, k$.

Thus, for $B \subset X$ we have

$$\lambda(B \times Y) = \lambda\left(\bigcup_{j=1}^{k} (B \cap A_j) \times Y\right) = \sum_{j=1}^{k} \lambda((B \cap A_j) \times Y) = \sum_{j=1}^{k} \mu(B \cap A_j) = \mu(B),$$

and for $C \subset Y$ we have

$$\lambda(X \times C) = \lambda\left(\bigcup_{j=1}^{k} X \times (C \cap \overline{\theta}(A_j))\right) = \sum_{j=1}^{k} \lambda(X \times (C \cap \overline{\theta}(A_j)))$$

$$= \sum_{j=1}^{k} \nu(C \cap \overline{\theta}(A_j)) = \nu(C).$$

Therefore the marginals of λ are μ and ν which implies $\lambda \in J(\mathbf{X}, \mathbf{Y})$. Furthermore,

$$\lambda(\{(x, y) : h(x, y) < \varepsilon\}) = \sum_{j=1}^{k} \lambda(A_j \times \{y \in \overline{\theta}(A_j) : h(x, y) < \varepsilon, x \in A_j\})$$

$$= \sum_{j=1}^{k} \nu(\{y \in \overline{\theta}(A_j) : h(x, y) < \varepsilon, x \in A_j\})$$

$$= \nu(\{y = \overline{\theta}(x) : h(x, y) < \varepsilon\})$$

$$\geq \nu(\overline{\theta}(X \setminus (E_1 \cup E_2 \cup \{x \in X : \overline{\theta}(x) \neq \theta(x)\}))).$$

(3.5)

Set $V := X \setminus (E_1 \cup E_2 \cup \{x \in X : \overline{\theta}(x) \neq \theta(x)\})$. Again by (ii) we have

$$\nu(\overline{\theta}(V)) > (1 - \varepsilon)\mu(X \setminus (E_1 \cup E_2 \cup \{x \in X : \overline{\theta}(x) \neq \theta(x)\}))$$

$$\geq (1 - \varepsilon)(\mu(E_1) - \mu(E_2) - \mu(\{x \in X : \overline{\theta}(x) \neq \theta(x)\}))$$

(3.6)

$$> (1 - \varepsilon)(1 - (C + 2)\varepsilon)$$

$$> 1 - (C + 3)\varepsilon.$$

Thus, taking $D := C + 3$, by (3.5) and (3.6) we have

$$\lambda(\{(x, y) : h(x, y) < \varepsilon\}) \geq 1 - D \cdot \varepsilon$$

as we wanted. □

3.1.2   Very Weak Bernoulli Partitions and Ornstein Theorems

As defined in Chap. 1, a measure preserving automorphism $f : X \rightarrow X$ is said to be a Bernoulli automorphism if it is measurably isomorphic to some Bernoulli shift. Finding such isomorphism, or proving it does not exist in a general situation is not an easy task. In Sect. 2.7.2 the Theorem 2.19 provided an alternative way of obtaining the Bernoulli property, namely one may prove that an automorphism f is Bernoulli by finding a Bernoulli partition (see Definition 2.17) for f. It turns out that even with this alternative way it is not easy to directly construct Bernoulli partitions. What comes to rescue is then a beautiful and deep theory developed by Ornstein nowadays called Ornstein Theory. By using the $\overline{d}$-distance in the space of finite partitions, Ornstein developed a way to prove that certain automorphisms are Bernoulli without properly finding a Bernoulli partition directly, but instead, by finding partitions which are "almost Bernoulli." The main goal of this section is to introduce the concept of Very Weak Bernoulli partitions and present two theorems due to Ornstein which will be crucial to develop the strategy to obtain Bernoulli property for the ergodic automorphisms which we are interested in.

Definition 3.6 Let $f : X \rightarrow X$ be a μ-preserving isomorphism of a measure space (X, μ). A partition α of X is called a Very Weak Bernoulli partition (VWB) for f if for any $\varepsilon > 0$ there exists $N_0 = N_0(\varepsilon)$ such that for any $N' \geq N \geq N_0, n \geq 0$, and ε-almost every element $A \in \bigvee_{k=N}^{N'} f^k \alpha$, we have

$$\overline{d}(\{f^{-i}\alpha\}_1^n, \{f^{-i}\alpha|A\}_1^n) \leq \varepsilon,$$

where the partition $\alpha|A$ is considered with the normalized measure $\mu/\mu(A)$.

Theorem 3.1 *If α is a Bernoulli partition for an automorphism f, then α is a Very Weak Bernoulli partition for f.*

Proof Let α be a Bernoulli partition. Take any $N' \geq N \geq 1$ and $A \in \bigvee_N^{N'} f^k \alpha$ and call

$$\bigvee_N^{N'} f^k \alpha = \{A_1, A_2, \ldots, A_l\}.$$

We will prove that $\overline{d}(\{f^{-i}\alpha\}_1^n, \{f^{-i}\alpha|A\}_1^n) = 0$ for any $n \geq 1$ and, consequently we will conclude that α will be VWB. The idea is to apply Lemma 3.2 to estimate the $\overline{d}$-distance.

For each $1 \leq i \leq l$, since the pairs $(A_i, \mu/\mu(A_i))$ and $(A_i \cap A, \mu/\mu(A_i \cap A))$ are Lebesgue spaces with the σ-algebras induced from the σ-algebra $\mathscr{A}$ of X, we can take a (normalized) measure preserving isomorphism $\theta_i : A_i \rightarrow A_i \cap A$. Let $\theta : X \rightarrow A$ be the isomorphism given by

$$\theta(x) = \theta_i(x), \quad \text{for } x \in A_i.$$

In particular, for any $x \in X$, if $x \in f^{-i}(E)$, with $E \in \alpha$, then $\theta(x) \in f^{-i}(E)$ which implies that the names of x and $h(x)$, with respect to $\{f^{-i}\alpha\}_1^n$ and $\{f^{-i}\alpha|A\}_1^n$ respectively, are the same. Thus $h(x, \theta(x)) = 0$ for every $x \in X$. In particular, θ satisfies the first item of Lemma 3.2 for any $\varepsilon > 0$.

Let us prove that θ also satisfies the second item of Lemma 3.2 for any $\varepsilon > 0$ by proving that θ is actually (normalized) measure preserving. As α is a Bernoulli partition it is generating, thus it is enough to prove that θ preserves the measure of each $B \in \bigcup_{j=-\infty}^{\infty} f^j \alpha$.

Let $B \in \bigcup_{j=-\infty}^{\infty} f^j \alpha$, say $B = f^k(E)$ for a certain $k \in \mathbb{Z}$ and $E \in \alpha$. Then, since θ_j is measure preserving for $1 \le j \le l$, we have

$$\frac{\mu(\theta(B))/\mu(A)}{\mu(B)} = \frac{1}{\mu(B)\mu(A)} \sum_{j=1}^{l} \mu(\theta_j(B \cap A_j))$$

$$= \frac{1}{\mu(B)\mu(A)} \sum_{j=1}^{l} \frac{\mu(A_j \cap A)\mu(B \cap A_j)}{\mu(A_j)}. \tag{3.7}$$

Now, as α is a Bernoulli partition $\{f^n\alpha\}_{n \ge 1}$ is an independent sequence of partitions which implies that $\mu(A_j \cap A) = \mu(A_j) \cdot \mu(A)$ and $\mu(B \cap A_j) = \mu(B) \cdot \mu(A_j)$ (remember that $B \in f^k(\alpha)$, $k \in \mathbb{Z}$). Substituting in (3.7) we obtain

$$\frac{\mu(\theta(B))/\mu(A)}{\mu(B)} = \frac{1}{\mu(B)\mu(A)} \sum_{j=1}^{l} \frac{\mu(A_j)\mu(A)\mu(B)\mu(A_j)}{\mu(A_j)}$$

$$= \sum_{j=1}^{l} \mu(A_j) = 1.$$

Thus θ satisfies the second item of Lemma 3.2 for any $\varepsilon > 0$. In particular, by the referred lemma, for every $\varepsilon > 0$ we obtain

$$\bar{d}(\{f^{-i}\alpha\}_1^n, \{f^{-i}\alpha|A\}_1^n) \le c \cdot \varepsilon$$

for a certain constant c. Thus $\bar{d}(\{f^{-i}\alpha\}_1^n, \{f^{-i}\alpha|A\}_1^n) = 0$ for any $n \ge 1$ as we wanted to show. $\qquad\square$

The following theorems of Ornstein relate the existence of VWB partitions with the Bernoulli property for a certain system.

Theorem 3.2 ([6]) *If α is a VWB partition of X, then*

$$\left(X, \bigvee_{-\infty}^{+\infty} f^{-i}\alpha, \mu, f \right)$$

is a Bernoulli system.

Theorem 3.3 ([5]) *If $\alpha_1 < \alpha_2 < \ldots$ is an increasing sequence of partitions of a Lebesgue space $(X, \mathscr{A}, \mu)$ such that*

$$\bigvee_{i=1}^{+\infty} \bigvee_{n=-\infty}^{+\infty} f^{-n}\alpha_i = \mathscr{A}$$

and, for each i, $(X, \bigvee_{n=-\infty}^{+\infty} f^{-n}\alpha_i, \mu, f)$ is a Bernoulli system, then $(X, \mathscr{A}, \mu, f)$ is a Bernoulli system.

Remark 3.2 A direct consequence of Theorems 3.2 and 3.3 is that if $\alpha_1 < \alpha_2 < \ldots < \alpha_n < \ldots$ is an increasing sequence of VWB partitions such that $\bigvee_{i=1}^{\infty} \alpha_i = \mathscr{A}$ then $(X, \mathscr{A}, \mu, f)$ is a Bernoulli system. This observation will be useful to prove Bernoulli property for ergodic automorphisms of $\mathbb{T}^2$ and for Anosov diffeomorphisms.

3.2 Ergodic Automorphisms of $\mathbb{T}^2$ Are Kolmogorov

Recall that between the mixing and the Bernoulli property we have the so-called Kolmogorov property. To prove that ergodic automorphisms f of $\mathbb{T}^2$ are Bernoulli one shows first that f is Kolmogorov and then, combining the Kolmogorov property with the geometric invariant structure of f, one obtains the Bernoulli property.

The Kolmogorov property for Ergodic automorphisms of $\mathbb{T}^2$ will be presented here as a direct consequence of a much more general result. More precisely the Kolmogorov property for ergodic automorphisms of $\mathbb{T}^2$ is a direct consequence from the fact that these automorphisms are Anosov diffeomorphisms (see Definition 4.1 and Example 4.1) and from Theorems 4.7 and 4.8.

Theorem 3.4 *Every ergodic automorphism of $\mathbb{T}^2$ is a K-automorphism.*

Proof If $f : \mathbb{T}^2 \to \mathbb{T}^2$ is an ergodic automorphism, then it has no roots of unity as eigenvalues. Now, as f is invertible it is also volume preserving. Consequently the integer matrix L which induces f has two eigenvalues λ_1, λ_2 with $|\lambda_1| < 1 < |\lambda_2|$ from where we conclude that f is an Anosov diffeomorphism. By Theorems 4.7 and 4.8 it follows that f is a Kolmogorov automorphism.   □

We remark that Theorem 3.4 could also be obtained as a direct consequence of the following result of Juzvinskii, which is a generalization of a result of Rohlin [8] who proved the same result with the additional hypothesis that the group was abelian.

Theorem 3.5 (Juzvinskii [4]) *Let G be a compact topological group and $f : G \to G$ an ergodic (with respect to the Haar measure) automorphism, then f is a Kolmogorov automorphism.*

3.3 Ergodic Automorphisms of $\mathbb{T}^2$ Are Bernoulli

We are now able to prove the main result of this chapter.

Theorem 3.6 ([7]) *If $A : \mathbb{T}^2 \to \mathbb{T}^2$ is an ergodic automorphism, then A is Bernoulli.*

The argument we present here to prove Theorem 3.6 is the same argument originally given by Ornstein and Weiss in [7] and the proof is divided in several steps.

As observed in Remark 3.2, by Theorems 3.2 and 3.3 to prove that an ergodic automorphism $f : \mathbb{T}^2 \to \mathbb{T}^2$ is Bernoulli it is enough to find a sequence of VWB partitions $\alpha_1 < \alpha < 2 \ldots$ such that $\bigvee_{i=1}^{\infty} \alpha_i = \mathscr{B}$ where $\mathscr{B}$ is the Borel σ-algebra of $\mathbb{T}^2$. Observe that if we take $\widetilde{\alpha_i}$ the standard partition of $[0, 1]^2$ by 2^{2i} rectangles whose sizes have length $1/2^i$ then $\{\alpha_n\}_{n\in\mathbb{N}}$ is a decreasing sequence of partitions and it is not difficult to prove that

$$\bigvee_{i=1}^{\infty} \alpha_i = \mathscr{B}.$$

Thus, it is enough to prove that these partitions α_i, $i \in \mathbb{N}$, are VWB. We actually prove a more general fact. We will show that any finite partition of $\mathbb{T}^2$ whose atoms have piecewise smooth boundaries is a VWB partition for f.

3.3.1 Invariant Spaces and Rectangles

Let $f : \mathbb{T}^2 \to \mathbb{T}^2$ be a linear ergodic automorphism. As done in Example 2.1 f is induced by a 2×2 integer matrix, which we will denote by L, with determinant equal to one and without unitary eigenvalues. Denote by λ_s, λ_u the eigenvalues of L where

$$|\lambda_s| < 1 < |\lambda_u|.$$

Associated to these eigenvalues we have two lines through the origin on $\mathbb{R}^2$, F^s and F^u, respectively, which are L-invariant as subspaces of $\mathbb{R}^2$, that is,

$$L \cdot F^s = F^s, \quad L \cdot F^u = F^u.$$

Furthermore, since F^s and F^u are the eigenspaces associated to λ_s and λ_u, respectively, we have $Lv_s = \lambda_s v_s$ and $L\lambda_u = \lambda_u v_u$ for all $v_s \in F^s, v_u \in F^u$. Given $x \in \mathbb{R}^2$ denote the translations of F^s and F^u by the vector x by

$$F^s(x) = F^s + x, \quad F^u(x) = F^u + x.$$

If we denote by $|\cdot|$ the usual norm on $\mathbb{R}^2$, we have

$$|L(x) - L(y)| = |\lambda_s||x - y|, \quad \text{if } y \in F^s(x)$$

$$|L(x) - L(y)| = |\lambda_u||x - y|, \quad \text{if } y \in F^u(x).$$

Let $\pi : \mathbb{R}^2 \to \mathbb{T}^2$ be the natural projection over $\mathbb{T}^2$. We define the *stable* and *unstable* manifolds of f through x by

$$\mathscr{F}^s(x) = \pi \circ F^s(x) \quad \text{and} \quad \mathscr{F}^u(x) = \pi \circ F^u(x)$$

respectively.

Given a set $J \subset \mathbb{T}^2$ and a point $x \in \mathbb{T}^2$, for the sake of simplicity we will abuse notation and denote by $\mathscr{F}^\tau(x) \cap J$ the connected component of $\mathscr{F}^\tau(x) \cap J$ which contains x where $\tau \in \{s, u\}$.

Definition 3.7 A set $R \subset \mathbb{T}^2$ is called a δ-rectangle centered in a point $w \in \mathbb{T}^2$ (Fig. 3.3) if it satisfies the following conditions:

(i) $w \in R \subset B(w, \delta)$;
(ii) for any $x, y \in R$ we have

$$(\mathscr{F}^s(x) \cap R) \cap (\mathscr{F}^u(x) \cap R) = \{z\} \subset R.$$

Definition 3.8 Given a rectangle R and a subset $E \subset \mathbb{T}^2$, we say that the intersection $E \cap R$ is an u-tubular intersection if for every $x \in E \cap R$ we have

$$\mathscr{F}^u(x) \cap R \subset R \cap E.$$

Fig. 3.3 Representation of a rectangle R in $\mathbb{T}^2$ with respect to a certain linear ergodic automorphism $f : \mathbb{T}^2 \to \mathbb{T}^2$

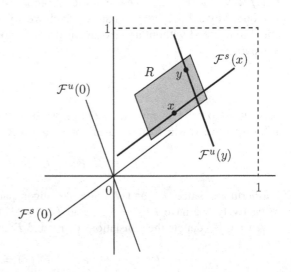

Definition 3.9 Let $\alpha = \{A_1, \ldots, A_k\}$ be a finite partition of $\mathbb{T}^2$. We say that α is a SB-partition (smooth boundary partition) if each A_i, $1 \leq i \leq k$, is a connected set with piecewise smooth boundary, that is, the boundary ∂A_i is a union of a finite number of smooth curves.

Lemma 3.3 *Let α be an SB-partition, R a rectangle, and $\delta > 0$ a given real number. There exists a sufficiently large natural number N_1 such that for any $N' > N \geq N_1$ and δ-almost every atom $P \in \bigvee_N^{N'} f^k \alpha$ there exists a subset $E \subset P$ with*

(i)

$$\frac{m(E)}{m(P)} > 1 - \delta;$$

(ii) the intersection $E \cap R$ is a u-tubular intersection.

Proof Observe that by the invariance of the stable and unstable leaves the tubularity of an intersection is not affected by applying f. Given an element $P \in \alpha$ denote by G_k^P the subset of non u-tubular intersections of R and $f^k(P)$, more specifically

$$G_k^P = \{y \in R \cap f^k(P) : \mathscr{F}^u(y) \cap R \not\subset R \cap f^k(P)\}.$$

We now estimate the measure of G_k^P.

Claim For a certain constant $C > 0$, given any $y \in f^{-k}(G_k^P)$ we have

$$d(y, \partial P) \leq C \cdot |\lambda_u|^{-k}. \tag{3.8}$$

Proof Consider $y \in f^{-k}(G_k^P)$ arbitrary. Since y is in the non-tubular intersection set we have that $\mathscr{F}^u(y) \cap f^{-k}(R) \not\subset f^{-k}(R) \cap P$. Since the connected component $\mathscr{F}^u(y) \cap f^{-k}(R)$ contains a point in P and a point w outside P, then $\mathscr{F}^u(y) \cap f^{-k}(R)$ intersects the boundary of P in a point z, in particular

$$d(y, w) \leq d(y, z) + d(z, w).$$

As the set R is a rectangle and the leaves $\mathscr{F}^u(x)$ are exponentially contracted with negative powers of f we have

$$d(y, w) \leq d(f^k(y), f^k(w)) \cdot |\lambda_u|^{-k} \leq \operatorname{diam}(R) \cdot |\lambda_u|^{-k}.$$

Thus it is enough to take $C = \operatorname{diam}(R)$.   △

Since the boundary of P is piecewise smooth, by (3.8) we have $m(G_k^P) \leq C \cdot |\lambda_u|^{-k}$. Since $|\lambda_u| > 1$ we can take N_1 large enough so that

$$m(G) \leq \sum_{k=N_1}^{+\infty} m(G_k^P) \leq \delta^2,$$

where $G := \bigcup_{k=N_1}^{+\infty} G_k^P$. Let

$$\Omega := \left\{ B \in \bigvee_N^{N'} f^k \alpha : \frac{m(B \cap G)}{m(B)} > \delta \right\}$$

and take $J := \bigcup_{B \in \Omega} B$. Assume that $m(J) > \delta$. Note that since J is a disjoint union we have

$$m(J) = \sum_{B \in \Omega} m(B).$$

Therefore

$$\delta^2 > m(G) \geq m(G \cap J) = \sum_{B \in \Omega} m(G \cap B) > \sum_{B \in \Omega} \delta \cdot m(B) = \delta \cdot m(J) > \delta^2,$$

contradiction. That is, δ- almost every atom $B \in \bigvee_N^{N'} f^k \alpha$ intersects G in a set with relative measure at most δ. Finally, for each atom $P \in \bigvee_N^{N'} f^k \alpha$ with $P \notin \Omega$, take $E = P \cap G^c$. E is u-tubular in R and since $P \notin \Omega$ we have (by definition of Ω)

$$\frac{m(P \cap G)}{m(P)} \leq \delta \Rightarrow \frac{m(P \cap G^c)}{m(P)} > 1 - \delta \Rightarrow \frac{m(E)}{m(P)} > 1 - \delta.$$

$\square$

3.3.2 Construction of the Function θ

In order to apply Lemma 3.2 to estimate the $\overline{d}$-distance, given a u-tubular subset $E \subset R$ we show how to construct a function $\theta : E \to R$ satisfying the hypothesis of Lemma 3.2. The classical Fubini's Theorem (see Theorem 2.6) plays a key role at this stage.

The next lemma shows how to construct the desired function θ (satisfying the hypothesis of Lemma 3.2) and may be considered as the most important step of the proof of Theorem 3.6.

Lemma 3.4 *Given $\delta_1 > 0$, there exists $\delta_2 > 0$ such that if R is a δ_2-rectangle and E is a u-tubular subset of R, then there exists a bijection $\theta : E \to R$ such that*

(1) θ is measure preserving;
(2) for every $k \geq 1$ and $x \in E$

$$d(f^k(\theta x), f^k(x)) < \delta_1.$$

Proof Since f contracts distances on $\mathscr{F}^s$ the second item will be satisfied once we take δ_2 small enough and guarantee that $\theta(x) \in \mathscr{F}^s(x)$.

Fix a point x_0 on the interior of R, $x_0 \in \text{Int}(R)$. Now consider the arcs $E \cap \mathscr{F}^s(x_0) \cap R$ and $\mathscr{F}^s(x_0) \cap R$. Define $\theta_0 : E \cap \mathscr{F}^s(x_0) \to \mathscr{F}^s(x_0) \cap R$ the standard linear map between these arcs mapping extreme points to extreme points. Observe that this map preserves the respective normalized Lebesgue measures of $E \cap \mathscr{F}^s(x_0)$ and $\mathscr{F}^s(x_0) \cap R$. Now we make use of u-tubularity of E with respect to R to construct θ as follows.

Take any $x \in R$. Define $\pi^u_{x,x_0} : \mathscr{F}^s(x) \cap R \to \mathscr{F}^s(x_0) \cap R$ by

$$\pi^u_{x,x_0}(y) = (\mathscr{F}^u(y) \cap R) \cap (\mathscr{F}^s(x_0) \cap R).$$

This function is called the unstable holonomy from $\mathscr{F}^s(x) \cap R$ to $F^s(x_0) \cap R$. Observe that π^u_{x,x_0} is nothing more than the projection of $\mathscr{F}^s(x)$ over $\mathscr{F}^s(x_0)$ parallel to $\mathscr{F}^u(x)$. In particular, it sends the Lebesgue measure λ^x_s of $\mathscr{F}^s(x)$ to the Lebesgue measure λ_s of $\mathscr{F}^s(x_0)$. By the u-tubularity of the set E inside R we have that π^u_{x,x_0} is also an isomorphism from $E \cap \mathscr{F}^s(x)$ to $E \cap \mathscr{F}^s(x_0)$ thus we also have

$$\lambda^x_s(R \cap \mathscr{F}^s(x)) = \lambda_s(\pi^u_{x,x_0}(R \cap \mathscr{F}^s(x))) = \lambda_s(R \cap \mathscr{F}^s(x_0)) \qquad (3.9)$$

and

$$\lambda^x_s(E \cap \mathscr{F}^s(x)) = \lambda_s(\pi^u_{x,x_0}(E \cap \mathscr{F}^s(x))) = \lambda_s(E \cap \mathscr{F}^s(x_0)). \qquad (3.10)$$

Finally define $\theta : E \to R$ as follows (see Fig. 3.4): for each $x \in E$, given $y \in \mathscr{F}^s(x) \cap R$ we define

Fig. 3.4 The measure preserving function $\theta : E \to R$

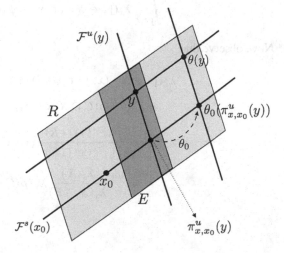

$$\theta(y) := (\pi^u_{x,x_0})^{-1} \circ \theta_0 \circ \pi^u_{x,x_0}(y).$$

Observe that this definition is only possible because E is u-tubular inside R. Also by (3.9) and (3.10), for $x \in R$, given any $B \subset \mathscr{F}^s(x) \cap E$ we have $\theta(B) \subset \mathscr{F}^s(x) \cap R$ and

$$\frac{\lambda^x_s(\theta(B))}{\lambda^x_s(\mathscr{F}^s(x) \cap R)} \overset{(3.9)}{=} \frac{\lambda^x_s((\pi^u_{x,x_0})^{-1}(\theta_0(\pi^u_{x,x_0}(B))))}{\lambda_s(X)} = \frac{\lambda_s(\theta_0(\pi^u_{x,x_0}(B)))}{\lambda_s(X)}$$

$$= \frac{\lambda_s(\pi^u_{x,x_0}(B))}{\lambda_s(X \cap E)} \overset{(3.10)}{=} \frac{\lambda^x_s(B)}{\lambda^x_s(E \cap \mathscr{F}^s(x))}, \tag{3.11}$$

where $X := \mathscr{F}^s(x_0) \cap R$. Let us prove that θ is measure preserving. Consider $Y = \mathscr{F}^u(x_0) \cap E = \mathscr{F}^u(x_0) \cap R$ and let λ_u be the Lebesgue measure on Y. Observe that by the definition of rectangle the set R may be identified with $X \times Y$ by the isomorphism

$$\rho : X \times Y \to R$$

$$(x, y) \mapsto \mathscr{F}^u(x) \cap \mathscr{F}^s(y)$$

which maps the measure $\frac{1}{\lambda_s(X)\lambda_u(Y)}\lambda_s \times \lambda_u$ to m. As E is u-tubular we have $\rho^{-1}(E) = (X \cap E) \times (Y \cap E)$. Consider $\psi : (X \cap E) \times (Y \cap E) \to X \times Y$ be defined by $\psi = \rho^{-1} \circ \theta \circ \rho$. Given any measurable set $C \subset (X \cap E) \times (Y \cap E)$, by Fubini's Theorem (Theorem 2.6) we have

$$\lambda_s \times \lambda_u(\psi(C)) = \int_{Y \cap E} \int_{X \cap E} \chi_{\psi(C)} d\lambda_s d\lambda_u$$

$$= \int_{Y \cap E} \lambda_s(\{x \in X : (x, y) \in \psi(C) \cap X \times \{y\}\}) d\lambda_u. \tag{3.12}$$

Now, observe that

$$\lambda_s(\{x \in X : (x, y) \in \psi(C) \cap X \times \{y\}\})$$

$$= \lambda^y_s(\theta \circ \rho(C \cap X \times \{y\}))$$

$$\overset{(3.11)}{=} \frac{\lambda^y_s(\mathscr{F}^s(y) \cap R)}{\lambda^y_s(\mathscr{F}^s(y) \cap E)} \lambda^y_s(\rho(C \cap X \times \{y\}))$$

$$\overset{(3.9),(3.10)}{=} \frac{\lambda_s(X)}{\lambda_s(X \cap E)} \lambda^y_s(\rho(C \cap X \times \{y\})),$$

which implies by (3.12)

$$\lambda_s \times \lambda_u(\psi(C)) = \int_{Y \cap E} \frac{\lambda_s(X)}{\lambda_s(X \cap E)} \lambda_s^y(\rho(C \cap X \times \{y\})) d\lambda_u$$

$$= \frac{\lambda_s(X)}{\lambda_s(X \cap E)} \int_{Y \cap E} \lambda_s(\{x \in X : (x, y) \in C \cap X \times \{y\}) d\lambda_u$$

$$= \frac{\lambda_s(X)}{\lambda_s(X \cap E)} \lambda_s \times \lambda_u(C).$$

As $\lambda_u(Y) = \lambda_u(Y \cap E)$ we conclude that ψ is measure preserving. Consequently θ is also measure preserving as we wanted to show. $\qquad\square$

Lemma 3.5 *Let $\varepsilon, \varepsilon' > 0$ and α be an SB-partition. Then, there exists $N_2 > 0$ such that for all $N' > N \geq N_2$ and for ε-almost every atom $P \in \bigvee_N^{N'} f^k \alpha$ there exists a subset $E \subset P$ and an injective function $\theta : E \to \mathbb{T}^2$ such that*

(1) θ is $c\varepsilon$- measure preserving for some constant $c > 0$;
(2)

$$\frac{m(E)}{m(P)} > 1 - \varepsilon;$$

(3) for every $k \geq 0$ and $x \in E$,

$$d(f^k(\theta x), f^k(x)) < \varepsilon'.$$

Proof Given $\varepsilon' > 0$, by applying Lemma 3.4 for $\delta_1 := \varepsilon'$ we obtain $\delta_2 > 0$ such that if R is a rectangle of diameter smaller than δ_2 and E is a u-tubular subset of R, then there exists an injective map $\theta : E \to R$ satisfying the properties (1) and (2) of Lemma 3.4.

Take a partition $\beta := \{R_0, \ldots, R_b\}$ where $R_1, \ldots, R_b$ are rectangles with diameter smaller than or equal δ_2 and $m(R_0) < \varepsilon/10$. Consider

$$\gamma := \varepsilon \cdot b^{-1} \cdot \min\{m(R_i) : 1 \leq i \leq b\}.$$

For each $1 \leq i \leq b$, let $N_1^i > 0$ be provided by Lemma 3.3 applied for $R := R_i$ and $\delta := \gamma$. Thus, taking $N_1 = \max\{N_1^i : 1 \leq i \leq b\}$ we have that, given $N' > N \geq N_1$, for ε-almost every atom P of $\bigvee_N^{N'} f^k \alpha$ and for each $1 \leq i \leq b$ there exists a subset $E_i \subset P$, constructed in the proof of Lemma 3.3, such that

$$\frac{m(E_i)}{m(P)} > 1 - \gamma \tag{3.13}$$

and $E_i \cap R_i$ is u-tubular in R_i for all $1 \leq i \leq b$. For such an atom P fixed let

$$E := \bigcup_{i=1}^{b} (E_i \cap R_i) \subset P. \tag{3.14}$$

It is easy to see that E intersects each R_i in a u-tubular subset and that $E \cap R_0 = \emptyset$ since $E \subset \bigcup_{i=1}^{b} R_i$. Furthermore, observe that by the construction of E_i (see Lemma 3.3) for $j \neq i$ we have $P \cap R_j = E_i \cap R_j$ since every point in $R_j \cap P$ is not in the non-tubular intersection with respect to R_i, thus we have by (3.13)

$$\frac{m(E)}{m(P)} = \sum_{i=1}^{b} \frac{m(E_i \cap R_i)}{m(P)} = \sum_{i=1}^{b} \frac{m(E_i) - \sum_{j \neq i} m(P \cap R_j)}{m(P)}$$

$$= \sum_{i=1}^{b} \frac{m(E_i)}{m(P)} - \sum_{i=1}^{b} \sum_{j \neq i} \frac{m(P \cap R_j)}{m(P)} \tag{3.15}$$

$$> b \cdot (1 - \gamma) - (b - 1) \sum_{j=1}^{b} \frac{m(P \cap R_j)}{m(P)}$$

$$> 1 - b\gamma > 1 - \varepsilon \cdot \min\{m(R_i) : 1 \leq i \leq b\}.$$

Also, for such $E \subset P$ we have

$$\left| \frac{m(E \cap R_i)}{m(E)} - m(R_i) \right| = \left| \left[\frac{m(P \cap R_i)}{m(P)} - m(R_i) \right] \frac{m(P)}{m(E)} + m(R_i) \left(\frac{m(P)}{m(E)} - 1 \right) \right|$$

$$< \frac{1}{1 - \varepsilon} \cdot \left| \frac{m(P \cap R_i)}{m(P)} - m(R_i) \right| + \frac{1}{1 - \varepsilon} - 1 \tag{3.16}$$

Now, by Theorem 3.4 f is Kolmogorov and, consequently, by Lemma 2.1 it follows that every finite partition, in particular the partition α, is Kolmogorov. That is, for each $1 \leq i \leq b$, given $\xi_i > 0$ there exists $N_0^i > 0$ such that for any $N' > N \geq N_0^i$ and ξ_i-almost every element $P \in \bigvee_{k=N}^{N'} f^k \alpha$ we have

$$\left| \frac{m(P \cap R_i)}{m(P)} - m(R_i) \right| \leq \xi_i. \tag{3.17}$$

Let $D > 0$ be a constant such that

$$D(1 - \varepsilon) \cdot \min\{m(R_i) : 1 \leq i \leq b\} > 1$$

and take

$$\xi_i < \varepsilon \cdot (D(1 - \varepsilon) \cdot \min\{m(R_i) : 1 \le i \le b\} - 1), \quad 1 \le i \le b,$$

and $N_2 := \max\{N_1, N_0^1, \ldots, N_0^b\}$. For ξ_i small enough, by (3.14)–(3.17) we have that for any $N' > N \ge N_2$ and ε-almost every element $P \in \bigvee_{k=N}^{N'} f^k \alpha$ there exists a subset $E \subset P$ intersecting each R_i in a u-tubular way satisfying

$$\frac{m(E)}{m(P)} > 1 - \varepsilon \cdot \min\{m(R_i) : 1 \le i \le b\} \ge 1 - \varepsilon \tag{3.18}$$

and, for each $1 \le i \le b$ we have

$$\left| \frac{m(E \cap R_i)}{m(E)} - m(R_i) \right|$$

$$< \frac{1}{1 - \varepsilon} \cdot \xi_i + \frac{1}{1 - \varepsilon} - 1$$

$$< \frac{1}{1 - \varepsilon}(\varepsilon \cdot (D(1 - \varepsilon) \cdot \min\{m(R_i) : 1 \le i \le b\} - 1) + 1) - 1$$

$$= D \cdot \varepsilon \cdot \min\{m(R_i) : 1 \le i \le b\} \le D \cdot \varepsilon \cdot m(R_i).$$

In particular we have

$$\left| \frac{m(E)m(R_i)}{m(E_i \cap R_i)} - 1 \right| \le \frac{D \cdot \varepsilon}{1 - D \cdot \varepsilon}, \quad 1 \le i \le b. \tag{3.19}$$

By Lemma 3.4, for each $1 \le i \le b$ we can define a measure preserving bijection $\theta_i : E_i \cap R_i \to R_i$ satisfying

$$d(f^k(\theta_i(x)), f^k(x)) < \varepsilon \tag{3.20}$$

for all $x \in E_i \cap R_i, k \ge 0$. Define the map $\theta : E \to \mathbb{T}^2$ by

$$\theta(x) = \theta_i(x) \quad \text{if } x \in E_i \cap R_i.$$

It is easy to see that θ is injective since each θ_i is bijective and the sets R_i are pairwise disjoint. Item (3) is clearly satisfied due to (3.20) and item (2) follows from (3.18). We are left to prove item (1).

Take $B \subset E$ a measurable set and denote $B_i := B \cap E_i \cap R_i = B \cap R_i$. Since $\theta_i : E_i \cap R_i \to R_i$ is measure preserving we have

$$\frac{m(\theta_i(B_i))}{m(R_i)} = \frac{m(B_i)}{m(E_i \cap R_i)},$$

which implies

$$\left| \frac{m(\theta_i(B_i))m(E)}{m(B_i)} - 1 \right| = \left| \frac{m(E)m(R_i)}{m(E_i \cap R_i)} - 1 \right|, \quad 1 \le i \le b. \tag{3.21}$$

Finally by (3.19) and (3.21) we obtain

$$\left| \frac{m(\theta(B))m(E)}{m(B)} - 1 \right| = \left| \frac{\sum_{i=1}^{b} m(\theta_i(B_i))m(E)}{m(B)} - 1 \right|$$

$$= \left| \sum_{i=1}^{b} \left[\frac{m(\theta_i(B_i))m(E)}{m(B_i)} - 1 \right] \cdot \frac{m(B_i)}{m(B)} \right|$$

$$\le \sum_{i=1}^{b} \frac{m(B_i)}{m(B)} \cdot \left| \frac{m(E)m(R_i)}{m(E_i \cap R_i)} - 1 \right|$$

$$\le \sum_{i=1}^{b} \frac{m(B_i)}{m(B)} \cdot \frac{D \cdot \varepsilon}{1 - D \cdot \varepsilon}$$

$$= \frac{D \cdot \varepsilon}{1 - D \cdot \varepsilon} < c \cdot \varepsilon,$$

for a certain constant $c > 0$. Thus, θ is $c\varepsilon$-measure preserving as we wanted to show.

$\square$

In the next lemma we prove that every SB-partition is VWB. Together with the observation made in the beginning of Sect. 3.3 (see also Remark 3.2) this result concludes the proof of Theorem 3.6.

Lemma 3.6 *If α is an SB-partition and $\varepsilon > 0$, then there exists N_2 such that for every $N' > N \ge N_2$ and ε-almost every atom $P \in \bigvee_N^{N'} f^k \alpha$ we have*

$$d(\{f^{-i}\alpha | P\}_1^n, \{f^{-i}\alpha\}_1^n) \le \varepsilon$$

for every $n \ge 1$. In other words, any SB-partition is VWB for A.

Proof Fix $\varepsilon > 0, 0 < \varepsilon' < \varepsilon$ and choose N_2 as in Lemma 3.5 and Y the exception set of atoms also obtained in Lemma 3.5, that is,

$$m(Y) \le c\varepsilon, \quad Y = \text{union of some elements of } \bigvee_{k=N}^{N'} f^k \alpha, \quad (N' > N \ge N_2).$$

Consider $P \in \bigvee_{k=N}^{N'} f^k \alpha$ with $P \cap Y = \emptyset$ and consider $E \subset P$ and $\theta : E \to \mathbb{T}^2$ be given by Lemma 3.5. Consider the sequence of partitions $\{\beta_i\}_1^n := \{f^{-i}\alpha\}_1^n$ and $\{\alpha_i\}_1^n := \{f^{-i}\alpha | P\}_1^n$, that is, each β_i is a partition of $\mathbb{T}^2$ and each α_i is a partition of P. Let $\{l_i(x)\}_{i=1}^n$ denote the α-name of the sequence of partitions $\{\alpha_i\}_1^n$ and $\{m_i(x)\}_{i=1}^n$ denote the β-name of the sequence of partitions $\{\beta_i\}_1^n$.

Now choose an arbitrary point $x_0 \in \mathbb{T}^2$ and define $\bar{\theta} : P \to \mathbb{T}^2$ by

$$\bar{\theta}(x) := \theta(x), \text{ for } x \in E$$

and

$$\bar{\theta}(x) = x_0 \text{ for } x \in P \setminus E.$$

We claim that $\bar{\theta}$ is $c_2 \cdot \varepsilon$-measure preserving for a certain constant $c_2 > 0$. Indeed, observe that

$$\frac{m(P \setminus E)}{m(P)} = 1 - \frac{m(E)}{m(P)} < 1 - (1 - \varepsilon) = \varepsilon.$$

Furthermore, if $B \subset E$ then $\bar{\theta}(B) = \theta(B)$ and then we have:

$$\left| \frac{m(\bar{\theta}(B))m(P)}{m(B)} - 1 \right| \leq \left| \frac{m(\bar{\theta}(B))m(E)}{m(B)} - 1 \right| \frac{m(P)}{m(E)} + \left| \frac{m(P)}{m(E)} - 1 \right| < \frac{c+1}{1-\varepsilon} \cdot \varepsilon.$$

Thus, taking $d > \max\left\{1, \frac{c+1}{1-\varepsilon}\right\}$ we have that $\bar{\theta}$ is $d\varepsilon$-measure preserving as we claimed.

Now, observe that by the third item of Lemma 3.5 we have

$$d(f^n(x), f^n(\bar{\theta}(x))) < \varepsilon'$$

for all $n \in \mathbb{N}$ and $x \in E$, since $\bar{\theta}$ coincides with θ in E. Thus, if $x \in E$, then

$$l_i(\theta(x)) \neq m_i(\theta(x))) = 1 \Rightarrow d(f^i(x), \partial \alpha_i^{l_i(x)}) < \varepsilon' \Rightarrow f^i(x) \in O_{\varepsilon'}(\partial \alpha_i^{l_i(x)}),$$
$$(3.22)$$

where $O_{\varepsilon'}(\partial \alpha_i^{l_i(x)})$ denotes the ε'-neighborhood of $\partial \alpha_i^{l_i(x)}$ and $\alpha_i^{l_i(x)}$ denotes the atom of α_i which contains x. Set

$$O_{\varepsilon'} := \bigcup_{A \in \bigcup_{i=1}^n \alpha_i} O_{\varepsilon'}(\partial A).$$

By (3.22) we have

$$h(x, \bar{\theta}(x)) = \frac{1}{n} \sum_{l_i(x) \neq m_i(\theta(x))} 1 \leq \frac{1}{n} \sum_{j=1}^n \chi_{O_{\varepsilon'}}(f^j(x)).$$

By ergodicity the right side converges to $m(O_{\varepsilon'})$ for almost every x. As the boundaries of atoms of A are piecewise smooth, $m(O_{\varepsilon'})$ can be taken arbitrarily small by choosing the initial ε' arbitrarily small. In particular, for ε' small, we have $m(O_{\varepsilon'}) < d \cdot \varepsilon$. Thus, $\overline{\theta}$ satisfies the hypothesis of Lemma 3.2 with ε replaced by $d\varepsilon$. Thus, there exists a constant $c_2 > 0$ such that

$$d(\{\alpha_i\}_1^n, \{\beta_i\}_1^n) \leq c_2\varepsilon,$$

for a certain constant $c_2 > 0$. Since $\varepsilon > 0$ is arbitrary it follows that α is VWB as we wanted to show.   □

This concludes the proof of Theorem 3.6.

References

1. Chernov, N.I., Haskell, C.: Nonuniformly hyperbolic K-systems are Bernoulli. Ergodic Theory Dyn. Syst. **16**, 19–44 (1996).
2. Glasner, E.: Ergodic Theory via Joinings. Mathematical Surveys and Monographs, vol. 101. American Mathematical Society, Providence (2003)
3. Halmos, P.R.: Measure Theory. D. Van Nostrand Company, New York (1950)
4. Juzvinskiĭ, S.A.: Metric properties of the endomorphisms of compact groups. Izv. Akad. Nauk SSSR Ser. Mat. **29**, 1295–1328 (1965)
5. Ornstein, D.S.: Two Bernoulli shifts with infinite entropy are isomorphic. Adv. Math. **5**, 339–348 (1970)
6. Ornstein, D.S.: Imbedding Bernoulli shifts in flows. In: Contributions to Ergodic Theory and Probability. Lecture Notes in Mathematics, vol. 160, pp. 178–218. Springer, Berlin (1970)
7. Ornstein, D.S., Weiss, B.: Geodesic flows are Bernoullian. Isr. J. Math. **14**, 184–198 (1973)
8. Rohlin, V.A.: Metric properties of endomorphisms of compact commutative groups. Izv. Akad. Nauk SSSR Ser. Mat. **28**, 867–874 (1964)

Chapter 4
Hyperbolic Structures and the Kolmogorov–Bernoulli Equivalence

Abstract In the previous chapter we have proved that linear ergodic automorphisms of $\mathbb{T}^2$ are Kolmogorov and, furthermore, they are Bernoulli. The main goal of this chapter is to show that Kolmogorov and Bernoulli property can be obtained for a much more general class of dynamical systems, namely those admitting a global uniform hyperbolic behavior, i.e., the Anosov systems (Definition 4.1). Anosov systems play a crucial role in smooth ergodic theory being the model for a huge variety of dynamical properties. In the first part of this chapter we make a quick introduction to the basic definitions and properties of uniformly hyperbolic systems and we will briefly present the geometric structures which are invariant by the dynamics of an Anosov diffeomorphism (Theorem 4.1). Linear ergodic automorphisms of $\mathbb{T}^2$ are very particular examples of Anosov diffeomorphisms. In light of this fact we will show how to obtain the Kolmogorov property for $C^{1+\alpha}$ Anosov diffeomorphisms (Theorem 4.8) and how we can use it to obtain the Bernoulli property (Theorem 4.9) in parallel to the argument used in Chap. 3.

In Sect. 4.1 we start by giving the definition of Anosov diffeomorphisms and some natural examples. Then we state the stable/unstable manifold theorem (Theorem 4.1) and show the main properties satisfied by the families of stable/unstable manifolds such as the absolute continuity property (Theorems 4.3 and 4.4). We finish Sect. 4.1 by using the so-called Hopf's argument to show that C^2 volume preserving Anosov diffeomorphisms are ergodic (Theorem 4.5). Using the results of Sect. 2.8 we then prove that C^2 volume preserving Anosov diffeomorphisms are Kolmogorov automorphisms (Theorem 4.8). In Sect. 4.2 we finally prove one of the main results of this book: C^2 volume preserving Anosov diffeomorphisms are Bernoulli (Theorem 4.9).

4.1 The Structure of Anosov Diffeomorphisms

One of the starting points in the study of the local behavior of a given dynamics $f : M \to M$ at a point $p \in M$ is to consider the set of points $y \in M$ for which $f^n(y)$ and $f^n(p)$ are close to each other either for every $n \geq 0$ (they are close in the whole future) or for every $n \leq 0$ (they are close in the whole past). To do so one defines the notion of *hyperbolic periodic point*: a periodic point $p \in M$ of period n is a hyperbolic periodic point for f if the linear map $(Df^n)_p : T_pM \to T_pM$ has no eigenvalue of modulus equal to one. The local dynamics around a hyperbolic periodic point is then proved to be highly influenced by the behavior of the linearized dynamics of f around p as showed by the classical Hartman-Grobman Theorem (see [4, Theorem 6.3.1]). In particular this shows that hyperbolic periodic points provide a good model for the study of local behavior of a certain dynamics. In light of this fact, the concept of a global hyperbolic structure, instead of just a hyperbolic periodic point, might be effective to study the global dynamical behavior in a similar manner that the one applied to the local analysis. Indeed this is what happens. In what follows we will introduce the concept of a hyperbolic set and an Anosov diffeomorphism and will show how the hyperbolic structure relates to the global dynamic behavior of the system. For a more complete introduction to the theory of Anosov diffeomorphisms and for the proof of theorems of this section, we refer the reader to [4].

Let M be a smooth manifold and $f : U \subset M \to M$ a C^1 diffeomorphism defined in an open set $U \subset M$. We say that a set $\Lambda \subset U$ is a compact f-invariant set if Λ is compact and $f(\Lambda) = \Lambda$.

Definition 4.1 A compact f-invariant set $\Lambda \subset M$ is called a hyperbolic set for f if there exists a Riemannian metric in an open neighborhood U of Λ and constants $\lambda < 1 < \mu$ such that for every $x \in \Lambda$ there exists a splitting

$$T_xM = E^s(x) \oplus E^u(x)$$

satisfying

1. the splitting is invariant , that is, $Df(x) \cdot E^s(x) = E^s(x)$ and $Df(x) \cdot E^u(x) = E^u(x)$;
2. for any vectors $v_s \in E^s(x)$, $v_u \in E^u(x)$ we have

$$||Df^n(x) \cdot v_s|| \leq \lambda^n ||v_s||, \quad \text{for all } n \in \mathbb{Z},$$
$$||Df^{-n}(x) \cdot v_u|| \leq \mu^{-n} ||v_u||, \quad \text{for all } n \in \mathbb{Z}. \tag{4.1}$$

If M is a hyperbolic set for a C^1 diffeomorphism $f : M \to M$, then we say that f is an Anosov diffeomorphism.

Example 4.1 As showed in Chap. 1, any matrix $L \in Sl(2, \mathbb{Z})$ induces a transformation $f : \mathbb{T}^2 \to \mathbb{T}^2$ and furthermore f is ergodic if, and only if, L has no eigenvalues of modulus 1. If f is ergodic, let λ^s and λ^u be the eigenvalues of L with $|\lambda^s| < 1 < |\lambda^u|$ and $E^s(x)$, $E^u(x)$ be the respective associate eigenspaces projected to the tangent space $T_x \mathbb{T}^2$. It is easy to see that

$$T_x \mathbb{T}^2 = E^s(x) \oplus E^u(x)$$

and (4.1) occurs for f. Thus every linear ergodic automorphism of $\mathbb{T}^2$ is an Anosov diffeomorphism.

Example 4.2 Given a C^1 Anosov diffeomorphism $f : M \to M$ of a compact Riemannian manifold M one can prove, by using an invariant-cones family characterization of the Anosov diffeomorphisms (see Section 6.2 from [4]), that C^1 perturbations of f are also Anosov (Corollary 6.4 in [4]). Thus, one can easily construct non-linear examples of Anosov diffeomorphisms in the following way. Let $f : \mathbb{T}^n \to \mathbb{T}^n$ be a linear Anosov automorphism and $h : \mathbb{T}^n \to \mathbb{T}^n$ be a non-linear C^1 diffeomorphism sufficiently C^1-close to identity. Then $h \circ f : \mathbb{T}^n \to \mathbb{T}^n$ is a C^1 non-linear Anosov diffeomorphism.

4.1.1 Invariant Foliations

Anosov diffeomorphisms are endowed with an extremely useful geometric structure. In the particular case of the linear ergodic automorphisms f of $\mathbb{T}^2$, we have seen in Sect. 3.3.1 that they admit two invariant structures, namely the stable and unstable manifolds. These manifolds are the projections of the eigenspaces of the linear matrix $L \in SL(2, \mathbb{Z})$ which induces f. The next theorem shows that any Anosov diffeomorphism admits a structure similar to the stable and unstable structure of the linear ergodic automorphisms of $\mathbb{T}^2$, more precisely, it shows that there exist two families of submanifolds which are invariant (as a family) and which admits asymptotic contracting/expanding behavior.

Theorem 4.1 (Stable and Unstable Manifolds Theorem [4]) *Let $f : M \to M$ be a C^1 Anosov diffeomorphism and let $\lambda < 1 < \mu$ be the numbers coming from (4.1). Then, for each $x \in M$ there is a pair of embedded C^1 discs $W^s(x)$, $W^u(x)$, called the local stable manifold and the local unstable manifolds of x, respectively, which satisfy the following:*

(1) $T_x W^s(x) = E^s(x)$ and $T_x W^u(x) = E^u(x)$;
(2) $f(W^s(x)) \subset W^s(f(x))$ and $f^{-1}(W^u(x)) \subset W^u(f^{-1}(x))$;
(3) for every $\epsilon > 0$ there exists $C(\epsilon)$ such that for all $n \in \mathbb{N}$ we have

$$d(f^n(x), f^n(y)) < C(\epsilon)(\lambda + \epsilon)^n d(x, y), \quad \text{for } y \in W^s(x),$$

$$d(f^{-n}(x), f^{-n}(y)) < C(\epsilon)(\mu - \epsilon)^{-n} d(x, y), \quad \text{for } y \in W^u(x),$$

where d is the distance on M induced by the Riemannian metric from the definition of Anosov diffeomorphism;

(4) there exists $\beta > 0$ and a family of neighborhoods O_x containing the ball around $x \in M$ of radius β such that

$$W^s(x) = \{y : f^n(y) \in O_{f^n(x)}, \quad n \in \mathbb{N}\},$$
$$W^u(x) = \{y : f^{-n}(y) \in O_{f^{-n}(x)}, \quad n \in \mathbb{N}\}.$$

Definition 4.2 We define the global stable manifold $\mathscr{F}^s(x)$ and the global unstable manifold $\mathscr{F}^u(x)$ of f at $x \in M$ by

$$\mathscr{F}^s(x) = \bigcup_{n=0}^{\infty} f^{-n}(W^s(f^n(x))),$$

$$\mathscr{F}^u(x) = \bigcup_{n=0}^{\infty} f^n(W^u(f^{-n}(x))).$$

It is not difficult to show that the global manifolds can be defined by the following topological characterization

$$\mathscr{F}^s(x) = \{y \in M : d(f^n(x), f^n(y)) \to 0 \text{ as } n \to \infty\}$$
$$\mathscr{F}^u(x) = \{y \in M : d(f^{-n}(x), f^{-n}(y)) \to 0 \text{ as } n \to \infty\}.$$

Remark 4.1 Although we omit the proof of Theorem 4.1 in the text we point out some important consequences of the construction of the local stable and local unstable manifolds of an Anosov diffeomorphism.

(1) The local stable (resp. local unstable) manifolds are not unique. However, it is easy to conclude from items (3) and (4) that if $W_1(x)$ and $W_2(x)$ are two local stable manifolds(resp. local unstable manifolds) satisfying (1)–(4) then $W_1(x) \cap W_2(x)$ contains an open neighborhood of x in $W_1(x)$ and an open neighborhood of x in $W_2(x)$.

(2) The local stable manifold $W^s(x)$ and the local unstable manifold $W^u(x)$ are transverse to each other at the point x since

$$T_x M = E^s(x) \oplus E^u(x) = T_x W^s(x) \oplus T_x W^u(x).$$

(3) The local stable manifolds $W^s(x)$ (resp. local unstable manifolds $W^u(x)$) depend continuously on x. Consequently, using the stability of transversal intersection of submanifolds we have that small perturbations of $W^s(x)$ and $W^u(x)$ are still transversal submanifolds. More precisely, there exists $\delta > 0$ such that for any $x, y \in M$ with $d(x, y) < \delta$ the intersection

$$W_\varepsilon^s(x) \cap W_\varepsilon^u(y) = \{[x, y]\}$$

is a unitary set, where $W_\varepsilon^s(x)$ and $W_\varepsilon^u(y)$ denote the ε-balls around x in $\mathscr{F}^s(x)$ and $\mathscr{F}^u(x)$, respectively.

(4) Consider a hyperbolic 2×2 matrix A, let v^u and v^s be the respective eigenvectors corresponding to eigenvalues greater than one and smaller than one, respectively. Notice that if we consider some vector v_0 transversal to the stable eigenspace, that is $v_0 = v_0^u + v_0^s$ where $v_0^u (\neq \emptyset)$ and v_0^s are in the unstable and stable eigenspace, respectively. Now observe that $A^n(v) = A^n(v_0^u) + A^n(v_0^s) = (\lambda^u)^n v_0^u + (\lambda^s)^n v_0^s$, hence the direction of $A^n(v)$ is tending towards the unstable direction, since $(\lambda^s)^n v_0^s$ tends to zero as n tends to infinity. This is exactly what happens in a slightly broader situation, if we consider a hyperbolic periodic point p and a transversal disc to the stable manifold, then if one iterated this disc by the dynamics, its iterates tend to the unstable manifold in the C^1 topology. This is a standard result and can be found on any smooth dynamical system book by the name of Inclination lemma or λ-lemma.

Theorem 4.2 (cf. Theorem 19.1.6 from [4]) *Let M be a compact Riemannian manifold and $f : M \to M$ a $C^{1+\alpha}$ Anosov diffeomorphism. Then the stable and unstable distributions are Hölder continuous.*

Definition 4.3 The family of global stable manifolds $\mathscr{F}^s = \{\mathscr{F}^s(x) : x \in M\}$ is called the stable foliation of f. Similarly, the family of global unstable manifolds $\mathscr{F}^u = \{\mathscr{F}^u(x) : x \in M\}$ is called the unstable foliation of f.

4.1.2 The Absolute Continuity Property of the Stable and Unstable Foliations

The fact that the stable and unstable foliations are not necessarily C^1 implies that we cannot apply the classical Fubbini's Theorem for these foliations. However, it is possible to show that the Jacobian of the stable holonomy is Hölder continuous and, as a consequence, we can show that the stable and unstable foliations are absolutely continuous. Before precisely stating the result let us recall the definition of absolute continuity and the concept of stable holonomy.

Definition 4.4 Let $T : X \to Y$ be an invertible measurable transformation between two measure spaces (X, ν) and (Y, μ). We say that G is absolutely continuous if $T_* \nu$ is absolutely continuous with respect to μ. In this case, we define the Jacobian of T at a point x to be the Radon–Nikodym derivative

$$\text{Jac}(T)(x) = \frac{d\mu}{dT_* \nu}.$$

If X is a metric space we have that for μ-almost every $x \in X$

$$\text{Jac}(T)(x) = \lim_{r \to 0} \frac{\mu(T(B(x, r)))}{\nu(B(x, r))}.$$

Let M be a compact Riemannian manifold and $f : M \to M$ be a $C^{1+\alpha}$ Anosov diffeomorphism. For fixed $x \in M, r > 0$, consider the family of local manifolds

$$\mathscr{L}(x) = \{W^u(w) : w \in B(x, r)\}$$

and choose two local disks, T^1, T^2 transverse to the family $\mathscr{L}(x)$.

Definition 4.5 The holonomy map between T_1 and T_2 generated by the local unstable manifolds, or the unstable holonomy map, is the function $\pi^u = \pi^u(x) :$ $D^1 \subset T_1 \to D^2 \subset T_2$, where D_1 is an open subset of T_1 and D_2 an open subset of T_2, defined by setting

$$\pi^u(y) = T^2 \cap W^u(w) \quad \text{if} \quad y = T^1 \cap W^u(w), \quad w \in B(x, r).$$

The next result establishes that the unstable (resp. stable) foliation is absolutely continuous and is one of the main results in smooth ergodic theory.

Theorem 4.3 ([2, 5]) *Let $f : M \to M$ be a $C^{1+\alpha}$ Anosov diffeomorphism of a compact Riemannian manifold M. Given T_1, T_2 two smooth local transversals to the local unstable (resp. stable) manifolds of f the unstable (resp. stable) holonomy π^u (resp. π^s), defined from an open subset of T_1 to an open subset of T_2, is absolutely continuous and the Jacobian $\mathrm{Jac}(\pi^u)$ (resp. $\mathrm{Jac}(\pi^s)$) is bounded from above and bounded away from zero.*

The proof of Theorem 4.3 is very technical and we postpone it to the next section in order to show some consequences of this theorem which are convenient at this point.

In general the partition by the leaves of the stable (resp. unstable) foliation $\mathscr{F}^s$ (resp. $\mathscr{F}^u$) may be non-measurable (recall Definition 2.6). Therefore, by disintegration of a measure along the leaves of such foliation we mean the disintegration on compact foliated boxes.

Definition 4.6 We say that the stable (resp. unstable) foliation $\mathscr{F}^s$ (resp. $\mathscr{F}^u$) is leafwise absolutely continuous if, for any $x \in M$ and $r > 0$ small enough, given ξ the partition of $B(x, r)$ induced by the leaves $\mathscr{F}^s(y)$, $y \in B(x, r)$, (i.e., the elements of ξ are connected components of $\mathscr{F}^s(y) \cap B(x, r)$) the conditional measures given by Rohklin's Theorem (see Theorem 2.7) along each element $L \in \xi$ are absolutely continuous with respect to the leaf volume measure in L.

Definition 4.7 We say that the stable (resp. unstable) foliation $\mathscr{F}^s$ (resp. $\mathscr{F}^u$) has atomic disintegration with respect to a measure μ, or that μ has atomic disintegration along $\mathscr{F}^s$ (resp. $\mathscr{F}^u$), if for any $x \in M$ and $r > 0$ small enough, given ξ the partition of $B(x, r)$ induced by the leaves $\mathscr{F}^s(y)$, $y \in B(x, r)$, (i.e., the elements of ξ are connected components of $\mathscr{F}^s(y) \cap B(x, r)$) the conditional measures given by Rohklin's Theorem along each element $L \in \xi$ are finite sums of Dirac measures.

Theorem 4.4 *Let M be a compact Riemannian manifold and $f : M \to M$ a $C^{1+\alpha}$ volume preserving Anosov diffeomorphism. Then the stable and unstable foliations are leafwise absolutely continuous.*

Proof Let $x \in M$ and $r > 0$. Consider ξ to be the partition of $B(x, r)$ given by the family of local stable manifolds $W^s(y)$, $y \in B(x, r)$. We denote by $m_{W^s}(w)$ the leaf volume measure on the local stable manifold $W^s(w)$ and by $m_s(w)$ the conditional measure on $W^s(w)$ obtained from the disintegration of the volume measure m on ξ.

Now, provided $r > 0$ is small enough we obtain a family of smooth local submanifolds $T(w)$, $w \in B(x, r)$ such that

(i) $T(w)$ is transversal to $W^s(w)$, $w \in B(x, r)$;
(ii) $T(w_1) \cap T(w_2) = \emptyset$ if $w_2 \notin T(w_1)$ and $T(w_1) = T(w_2)$ otherwise;
(iii) $B(x, r) \subset \bigcup_{w \in B(x,r)} T(w)$;
(iv) $T(w)$ depend smoothly on $w \in B(x, r)$.

We denote by $m_T(w)$ the leaf volume measure on $T(w)$. Let η be the partition of $B(x, r)$ generated by the family $\{T(w)\}_{w \in B(x,r)}$ and denote by $\{\tilde{\mu}(w)\}$ the conditional measures on $\{T(w)\}$ given by the disintegration of the volume measure m on the partition η, and $\hat{\mu}$ the factor measure on $B(x, r)/\eta$.

Let w_0 be any fixed point in $B(x, r)$ and consider

$$P(w_0) := \{y \in T(w_0) : \text{ there exists } z \in B(x, r) \text{ such that } y = T(w) \cap W^s(z)\}.$$

Given any $y \in P(w_0)$ we can identify the factor space $B(x, r)/\eta$ with $W^s(y)$. Now, since the local submanifolds $T(w)$ depend smoothly on w, we can apply the classical Fubbini's theorem to this family. In particular, by the Fubini's theorem there exists strictly positive smooth functions $g(w, z)$ and $h(y, w)$ such that

$$d\tilde{\mu}(w)(z) = g(w, z)dm_T(w)(z), \quad w \in B(x, r), z \in T(w) \tag{4.2}$$

and

$$d\hat{\mu}(w) = h(y, w)dm_{W^s}(y)(w), \quad y \in P(w_0). \tag{4.3}$$

Now, let $B \subset B(x, r)$ be a Borel set of positive measure. Then we have

$$m(B) = \int_{B(x,r)/\eta} \int_{T(w)} \chi_B(w, z)d\tilde{\mu}(w)(z)d\hat{\mu}(w)$$

$$\overset{(4.2)}{=} \int_{B(x,r)/\eta} \int_{T(w)} \chi_B(w, z)g(w, z)dm_T(w)(z)d\hat{\mu}(w)$$

$$= \int_{B(x,r)/\eta} \int_{T(w_0)} \chi_B(w, z)g(w, z) \operatorname{Jac}(\pi^s_{w_0,w})(y)dm_T(w_0)(y)d\hat{\mu}(w),$$

$$\tag{4.4}$$

where $\pi_{w_0,w}^s$ is the stable holonomy map between the transversals $T(w_0)$ and $T(w)$, and the point y is such that $z = \pi_{w_0,w}^s(y)$. Observe that to obtain the last equality we also used the fact that $\pi_{w_0,w}^s(P(w_0)) = P(w)$. Now, applying Fubbini's theorem to (4.4) we have that

$$m(B) = \int_{T(w_0)} \int_{B(x,r)/\eta} \chi_B(w,z)g(w,z)\,\mathrm{Jac}(\pi_{w_0,w}^s)(y)d\hat{\mu}(w)dm_T(w_0)(y)$$

$$\stackrel{(4.3)}{=} \int_{T(w_0)} \int_{W^s(y)} \chi_B(w,z)g(w,z)\,\mathrm{Jac}(\pi_{w_0,w}^s)(y)h(y,w)dm_{W^s}(y)(w)dm_T(w_0)(y).$$

Consequently, by the uniqueness of the disintegration, the conditional measure $m_s(y)$ on $W^s(y)$ is equivalent to the leaf volume measure $m_{W^s}(y)$ with density function

$$\rho^s(y,w) = \mathrm{Jac}(\pi_{w_0,w}^s)(y)g(w,\pi_{w_0,w}^s(y))h(y,w).$$

By Theorem 4.3 we know that $\mathrm{Jac}(\pi_{w_0,w}^s)(y)$ is bi-Hölder continuous in y and, consequently, it is positive and bounded. This implies that $m_s(y)$ is absolutely continuous with respect to the leaf volume measure $m_{W^s}(y)$ as we wanted to show.

$\square$

Corollary 4.1 *Let $B \subset M$ be a set of zero volume measure. Then*

$$m_{W^s}(x)(W^s(x) \cap B) = 0$$

for m-almost every point $x \in M$.

Proof Assume that there exists a set of zero volume B such that, for a certain set of positive volume $A \subset M$ we have

$$m_{W^s}(x)(W^s(x) \cap B) > 0, \quad \forall x \in A. \tag{4.5}$$

Let $x \in A$ be a density point of A. Now, consider the set

$$Q := \bigcup_{w \in B(x,r) \cap A} W^s(x) \cap B.$$

By Theorem 4.4 we have that the disintegration of m on the partition ξ, where ξ is the partition of $B(x,r)$ generated by the stable manifolds (see the proof of Theorem 4.4), is equivalent to the leaf volume measure. Thus for a certain (bounded and bounded away from zero) density function ρ we have

$$m(Q) = \int_{B(x,r)/\xi} \int_{W^s(w)} \rho(y,w)\chi_{W^s(w)\cap Q}(y)dm_{W^s}(w)(y)d\hat{m}(w),$$

where $\hat{m}$ is the factor measure on the transversal space $B(x,r)/\xi$. By (4.5) it follows that $m(Q) > 0$. But since $Q \subset B$ we have that $m(B) > 0$ which yields a contradiction. $\qquad\square$

Corollary 4.2 *Let $B \subset M$ be a measurable subset with $m(B) = 1$. Then for m-almost every point $x \in B$ we have*

$$m_{W^s}(x)(W^s(x) \setminus B) = 0.$$

Proof Assume by contradiction that we may find a full measure subset $B \subset M$ and a positive measure set $A \subset B$ such that

$$m_{W^s}(x)(W^s(x) \setminus B) > 0$$

for every $x \in A$. Then, by the same argument used in the proof of Corollary 4.1, the set $Q := \bigcup_{x \in A} W^s(x) \setminus B$ has positive volume. But $Q \subset M \setminus B$ implies $m(Q) = 0$ which contradicts the previous statement. $\qquad\square$

4.1.3   Proof of Theorem 4.3

The proof of Theorem 4.3 we present here is based on the argument of Pugh and Shub [5] adapted to the case of Anosov maps. The proof of the more general case considered in [5] is essentially the same but it is not convenient to treat the general case here as it involves several technicalities which are part of the general partially hyperbolic theory.

Let M be a smooth Riemannian manifold, $p \in M$ a point and $D^k \subset \mathbb{R}^k$ the unitary k-dimensional disc. The class of all C^r embeddings $\varphi : D^k \to M, r \geq 0$, for which $\varphi(0) = p$, is a metric space and will be denoted here by $\mathrm{Emb}^r(D^k, 0, M, p)$.

Definition 4.8 A pre-foliation of M by C^r D^k-discs is a map $p \mapsto D_p$, which associates to each $p \in M$ a C^r k-disc D_p in M containing p, and which is continuous in the following sense: for each $x \in M$, there exist an open set $x \in U_x$ and a continuous section $\sigma : U_x \to \mathrm{Emb}^r(D^k, U)$ such that:

(i) $M = \bigcup_{x \in M} U_x$;
(ii) $D_x = \sigma(x)(D^k), \quad x \in U_x.$

If, furthermore, these sections σ can all be chosen to satisfy:

(iii) the maps $(p, x) \mapsto \sigma(p)(x)$ are C^s, $1 \leq s \leq r$,

we say that the pre-foliation is of class C^s. For convenience, we also refer to the family of discs $\mathscr{D} = \{D_p\}_{p \in M}$ as being a pre-foliation.

Example 4.3 If N is a C^r sub-bundle of TM constituted of k-planes, then for $\delta > 0$ small enough the correspondence

$$p \longmapsto \exp_p(N_p(\delta)), \quad p \in M,$$

produces a C^r pre-foliation of M by C^∞ k-discs.

As for the case of foliations, pre-foliations naturally produce a family of maps between local transversals which we will call holonomy maps. Given a pre-foliation $\mathscr{G} = \{\mathscr{G}(p)\}_{p \in M}$ of M by C^∞ k-discs, let $\mathscr{G}(p)$ to denote the disc in $\mathscr{G}$ associated to p and let $q \in \operatorname{Int}\mathscr{G}(p)$. Let D_p and D_q two smooth $(m - k)$-embedded discs which are transverse to $\mathscr{G}(p)$ at p and w, respectively. In other words,

$$T_p D_p \oplus T_p\mathscr{G}(p) = T_p M, \quad T_q D_q \oplus T_q\mathscr{G}(p) = T_q M.$$

Then, as $\mathscr{G}(y)$ depends continuously on $y \in D_p$ and $\mathscr{G}(p)$ intersects D_q transversely at q, there exists an open neighborhood of p in D_p where we can define a surjective map $G_{p,q} : D_{p,q} \to R_{p,q} \subset D_q$ by sliding along the pre-foliation $\mathscr{G}$, that is, for $y \in D_{p,q}$

$$G_{p,q}(y) = \mathscr{G}(y) \cap D_q.$$

The maps $G_{p,q}$ are called the *holonomy maps*[1] defined by the pre-foliation $\mathscr{G}$. The map $G_{p,q}$ is C^s when $\mathscr{G}$ is of class C^s and it also depends continuously on p, q, D_p, D_q is the C^s sense. In particular, if $\mathscr{G}$ is C^1 and q and p are close enough, the map $G_{p,q}$ is a local diffeomorphism.

A key point in the proof is the following measure-theoretical result whose proof can be found in [5]. In what follows we use the double arrow notation $\rightrightarrows$ to denote uniform convergence of a sequence of functions.

Lemma 4.1 ([1, 5]) *Let $h : D^k \to \mathbb{R}^k$ to be a topological embedding and $\{g_n\}_{n \in \mathbb{N}}$ a sequence of C^1 embeddings, $g_n : D^k \to \mathbb{R}^k$ such that*

$$g_n \rightrightarrows h \quad and \quad J(g_n) \rightrightarrows J$$

where $J(g_n)$ denotes the Jacobian of g_n. Then h is absolutely continuous and its Jacobian is given by J.

By considering M as a manifold embedded into an euclidean space, we can define the angle between two subspaces of TM in the following manner: for A_p, B_p linear subspaces of $T_p M$,

$$\angle(A_p, B_p) = \max\{\angle(a, B_p) : a \in A_p \setminus \{0\}\} \cup \{\angle(b, A_p) : b \in B_p \setminus \{0\}\};$$

---

[1]In [5] these maps are called the Poincaré maps defined by the pre-foliation.

Now, the angle between two sub-bundles A and B of TM can be defined as

$$\measuredangle(A, B) := \sup_{p \in M} \measuredangle(A_p, B_p).$$

Lemma 4.2 *Assume that* $TM = N \oplus E^s = E^u \oplus E^s$ *where N is a smooth sub-bundle. Let $\mathscr{G}(\delta)$ be the smooth pre-foliation $p \to \mathscr{G}_p(\delta) = \exp_p(N_p(\delta))$. Given $0 \le \beta < \pi/2$, for $\delta > 0$ small enough, each holonomy map $G_{p,q} : D_{p,q} \to R_{p,q}$ along $\mathscr{G}(\delta)$ is an immersion if*

$$\measuredangle(TD_p, (E)^\perp) \le \beta \quad and \quad \measuredangle(TD_q, (E)^\perp) \le \beta.$$

Proof Since $G_{p,q}$ is smooth and has derivative continuously depending on p and q, it suffices to prove that $D_y G_{p,q} : T_y D_p \to T_{y'} D_q$ is a bijection where $y' := G_{p,q}(y)$. By definition of the holonomy map, if p is close enough to y, then $G_{p,q} = G_{y,y'}$, so that it is enough to verify the bijectivity at $y = p$. Observe that if $y = p = q$ then $G_{p,q}$ is the identity map, thus as the derivative of $G_{p,q}$ depends continuously on p, q, D_p, and D_q, and since $\{A_p \subset T_pM : \measuredangle(A_p, (E^u))\}$ is compact, for (p, q) in a neighborhood of the diagonal $\Delta = \{(p, p) : p \in M\}$ the map $D_{p,q}$ is also bijective concluding that $G_{p,q}$ is an immersion as we wanted. $\square$

Proof of Theorem 4.3 Let N be a smooth approximation of the unstable distribution E^u, that is, N is a smooth sub-bundle of TM whose angle to the unstable distribution can be chosen to be arbitrarily small. Consider $0 < \beta < \pi/2$ such that

$$\measuredangle(E^s, (E^u)^\perp) < \beta \quad e \quad \measuredangle(E^s, N^\perp) < \beta.$$

Now, let $\delta > 0$ be the value provided by Lemma 4.2 and consider $\mathscr{G}$ to be the pre-foliation given by

$$\mathscr{G} : \quad \mathscr{G}(y) := \exp_y(N_y(\delta)), \quad y \in M,$$

where $N_y(\delta)$ denotes the δ-ball in N_y centered at the origin. For each $n \in \mathbb{Z}$ the pre-foliation $\mathscr{G}$ naturally induces a pre-foliation $\mathscr{G}^n$ given by

$$\mathscr{G}^n : \quad \mathscr{G}^n(y) := f^n \mathscr{G}(f^{-n}(y)).$$

Also, denote by $\mathscr{G}^n(\varepsilon)$ the restriction of the elements of $\mathscr{G}^n$ to balls of radius ε, i.e.,

$$\mathscr{G}^n(\varepsilon) : \quad \mathscr{G}^n(y, \varepsilon) := \{x \in \mathscr{G}^n(y) : d_{\mathscr{G}^n}(x, y) \le \varepsilon\}.$$

By item (4) of Remark 4.1 we have $\mathscr{G}^n(\varepsilon) \rightrightarrows W^u(\varepsilon)$ e $T\mathscr{G}^n(\varepsilon) \rightrightarrows E^u$. Let $p \in M, q \in W^u(p)$ and two discs D_p and D_q which are transversal to E^u. Consider $H_{p,q} : D_{p,q} \to R_{p,q}$ to be the holonomy map associated to the foliation $\mathscr{F}^u$, where $D_{p,q}$ is a neighborhood of p in D_p and $R_{p,q}$ is a neighborhood of q in D_q.

The f-invariance of the unstable foliation yields

$$f^n \circ H_{f^{-n}(p), f^{-n}(q)} = H_{p,q} \circ f^n, \quad n \in \mathbb{Z}.$$

In particular, since f is a diffeomorphism, $H_{p,q}$ has a continuous positive Jacobian if and only if the same is true for $H_{f^{-n}(p), f^{-n}(q)}$.

Again by item (4) of Remark 4.1 we have $T(f^{-n}(D_p)), T(f^{-n}(D_q)) \rightrightarrows E^s$ as $n \to \infty$, thus we can assume without loss of generality that for all $n \geq 0$ we have

$$q \in W_p^u(\epsilon/2) \measuredangle(T(f^{-n}(D_p)), (E^u)^\perp) \leq \beta \text{ and } \measuredangle(T(f^{-n}(D_q)), (E^u)^\perp) \leq \beta. \tag{4.6}$$

Also, since the existence of the Jacobian is a local issue, we may assume that $D_p = D_{p,q}$ and $R_{p,q}$ is interior to D_q. Consider $g_n = G_{p,Q_n}^n \restriction D_p$, $Q_n = \mathscr{G}_p^n(\epsilon) \cap D_q$ and $h = H_{p,q}$. Note that $g_n \rightrightarrows h$ since $\mathscr{G}^n(\epsilon) \rightrightarrows \mathscr{F}^u$. We claim that

1. g_n is an embedding and
2. $J(g_n) \rightrightarrows J := \lim_{n \to \infty} \dfrac{\det(D_y f^{-n} \restriction_{T_y D_p})}{\det(D_{h(y)} f^{-n} \restriction_{T_{h(y)} D_q})}.$

Let us prove the first claim. Consider $\delta > 0$ as in Lemma 4.2 and β as in (4.6) so that each g_n in an immersion and, furthermore, observe that g_n and h can be defined in a disc $\widehat{D}_p$ slightly larger than D_p. Denote by $\widehat{g}_n$ and $\widehat{h}$ the extensions of g_n and h, respectively.

From the continuity and the uniform convergence of g_n we have $\widehat{g}_n \rightrightarrows \widehat{h}$. As M is compact, $\widehat{g}_n$ is locally injective and $\widehat{h}$ is a homeomorphism for every $y \in D_q$, we have that $\widehat{h}^{-1}(y)$ and $\widehat{g}_n^{-1}(y)$ are both finite. Let Y be a compact neighborhood of $R_{p,q} = h D_p$ inside the interior of $\widehat{h} \widehat{D}_p$. For every $y \in Y$, the degree of $\widehat{h}$ at y is $\deg(\widehat{h}, y) = 1$ since $\widehat{h}$ is a homeomorphism. Now, for n large the map $\widehat{g}_n \restriction_\partial \widehat{D}_p$ is arbitrarily close to $\widehat{h} \restriction_{\partial \widehat{D}_p}$, and consequently these maps are homotopic, i.e.,

$$\widehat{g}_n \restriction_\partial \widehat{D}_p \simeq \widehat{h} \restriction_{\partial \widehat{D}_p}.$$

In particular, by the invariance of the degree via homotopy we obtain

$$\deg(\widehat{g}_n, \partial \widehat{D}_p, y) = \deg(\widehat{h}, \partial \widehat{D}_p, y)$$

Now, since $\widehat{g}_n$ is an immersion for every $y \in Y$, $D_y \widehat{g}_n : T_y D_p \to T_{\widehat{g}_n y} D_q$ is injective and consequently also surjective, implying that every $y \in Y$ is a regular value of $\widehat{g}_n$. Thus for $y_1, y_2 \in Y$,

$$\deg(\widehat{g}_n, y_1) = \deg(\widehat{h}, y_2) \Rightarrow \deg(\widehat{g}_n, \widehat{D}_p, y) = 1,$$

for all $y \in Y$, which shows that $\widehat{g}_n$ is injective since it is an immersion. Thus $\widehat{g}_n$ embeds $\widehat{g}_n^{-1}(Y)$ and consequently g_n embeds D_p.

Now, let us prove the second claim. Note that we can write

$$g_n = f^n \circ G^0_{p_n,q_n} \circ f^{-n};$$

where $p_n = f^{-n}(p)$, $q_n = f^{-n}(q)$ and $Q_n = \mathscr{G}^n_p(\epsilon) \cap D_q$, and consequently for any $n \in N$, $f^{-n}(Q_n) \in \mathscr{G}_{f^{-n}(p)} = \mathscr{G}_{p_n}$, thus the pre-holonomy along $\mathscr{G}$, G_{P_n,q_n}, is well defined in $f^{-n}(D_p)$. Moreover, $T\mathscr{G}^n_p(\epsilon) \to E^u_p$ and $\mathscr{G}^n_p(\epsilon) \to W^u_p(\epsilon)$, thus,

$$q_n \in \mathscr{G}_{p_n}(\epsilon_n), \quad \text{when} \quad \epsilon_n \to 0, \ n \to \infty.$$

By the chain rule we have

$$D_y g_n = D_{G^0_{p_n,q_n}}(f^{-n}(y)) f^n \cdot D_{f^{-n}(y)} G^0_{p_n,q_n} \cdot D_y f^{-n}.$$

Thus,

$$J_y(g_n) = \det(D_y f^n \restriction_{T_{f^{-n}(g_n(y))} f^{-n}(D_q)}) \cdot \det(D_{f^n(y)} G^0_{p_n,q_n} \restriction_{T_{f^{-n}(y)} f^{-n}(D_p)})$$
$$\cdot \det(D_y f^{-n} \restriction_{T_y D_p}).$$

Now, since $\det(DG^0_{p_n,q_n} \restriction_{T_{f^{-n}(D_p)}}) \rightrightarrows 1$, to show the second claim is enough to show that

$$\lim_{n \to \infty} \frac{\det(D_y f^{-n} \restriction_{T_y D_p})}{\det(D_{g_n(y)} f^{-n} \restriction_{T_{g_n(y)} D_q})} = \lim_{n \to \infty} \frac{\det(D_y f^{-n} \restriction_{T_y D_p})}{\det(D_{h(y)} f^{-n} \restriction_{T_{h(y)} D_q})}.$$

Consider the case $y = p$, $T_p D_p = E^s_p$ and $T_q D_q = E^s_q$. Let us prove that for this case the limit

$$\lim_{n \to \infty} \frac{\det(D_p f^{-n} \restriction_{E^s_p})}{\det(D_q f^{-n} \restriction_{E^s_q})} \tag{4.7}$$

exists uniformly.

By the chain rule (4.7) is equivalent to the uniform convergence of

$$\prod_{k=0}^{\infty} \frac{\det(D_{f^{-k}(p)} f^{-1} \restriction_{E^s_{f^{-k}(p)}})}{\det(D_{f^{-k}(q)} f^{-1} \restriction_{E^s_{f^{-k}(q)}})}. \tag{4.8}$$

Lemma 4.3 *The uniform convergence of (4.8) holds if the series*

$$\sum_{k=0}^{\infty} |\det(D_{f^{-k}(p)} f^{-1} \restriction_{E^s_{f^{-k}(p)}}) - \det(D_{f^{-k}(q)} f^{-1} \restriction_{E^s_{f^{-k}(q)}})|. \tag{4.9}$$

is uniformly convergent.

Proof of Theorem 4.3 For any $n \in \mathbb{N}$ we have

$$\prod_{k=0}^{n} \frac{\det \left(D_{f^{-k}(p)} f^{-1} \restriction E^s_{f^{-k}(p)} \right)}{\det \left(D_{f^{-k}(q)} f^{-1} \restriction E^s_{f^{-k}(q)} \right)} = \exp \left(\sum_{k=0}^{n} \log \frac{\det \left(D_{f^{-k}(p)} f^{-1} \restriction E^s_{f^{-k}(p)} \right)}{\det \left(D_{f^{-k}(q)} f^{-1} \restriction E^s_{f^{-k}(q)} \right)} \right)$$

$$\leq \exp \left(\sum_{k=0}^{n} \left(\frac{\det \left(D_{f^{-k}(p)} f^{-1} \restriction E^s_{f^{-k}(p)} \right)}{\det \left(D_{f^{-k}(q)} f^{-1} \restriction E^s_{f^{-k}(q)} \right)} - 1 \right) \right),$$

where the last inequality comes from the fact that $\log x \leq x - 1$ for $x > 0$.

As f is a diffeomorphism we may assume that the determinants appearing in the right side are positive. Since E^s is a continuous distribution, f is a C^2 diffeomorphism and M is compact there exists a constant $C_1 > 1$ for which

$$C_1^{-1} \leq \det \left(D_{f^{-k}(q)} f^{-1} \restriction E^s_{f^{-k}(q)} \right) \leq C_1.$$

Thus

$$L \leq \exp \left(C_1 \sum_{k=0}^{\infty} \left| \det \left(D_{f^{-k}(p)} f^{-1} \restriction E^s_{f^{-k}(p)} \right) - \det \left(D_{f^{-k}(q)} f^{-1} \restriction E^s_{f^{-k}(q)} \right) \right| \right),$$

where $L := \prod_{k=0}^{n} \dfrac{\det \left(D_{f^{-k}(p)} f^{-1} \restriction E^s_{f^{-k}(p)} \right)}{\det \left(D_{f^{-k}(q)} f^{-1} \restriction E^s_{f^{-k}(q)} \right)}$, proving the statement of the lemma.   □

We are left to prove the convergence of (4.9). Since the distribution E^s is Hölder continuous and f is a C^2 diffeomorphism, there exists $D_2, \theta \in \mathbb{R}$ such that

$$\left| \det \left(D_{f^{-k}(q)} f^{-1} \restriction E^s_{f^{-k}(p)} \right) - \det \left(D_{f^{-k}(q)} f^{-1} \restriction E^s_{f^{-k}(q)} \right) \right|$$

$$\leq D_2 \cdot d \left(f^{-k}(p), f^{-k}(q) \right)^{\theta}.$$

Therefore, since $q \in W^u_p(\epsilon/2)$ we have

$$\left| \det \left(D_{f^{-k}(p)} f^{-1} \restriction E^s_{f^{-k}(p)} \right) - \det \left(D_{f^{-k}(q)} f^{-1} \restriction E^s_{f^{-k}(q)} \right) \right|$$

$$\leq D_2 \cdot d \left(f^{-k}(p), f^{-k}(q) \right)^{\theta}$$

$$\leq D_2 \cdot \mu^{-\theta k} \cdot d(p, q)^{\theta}.$$

Since $\mu > 1$ the series converges uniformly. That is, we have shown so far that the limit (4.7) exists uniformly.

Let us prove that the existence of the uniform limit (4.7) implies

$$\lim_{n \to \infty} \frac{\det \left(D_y f^{-n} \restriction_{T_y D_p} \right)}{\det \left(D_{g_n(y)} f^{-n} \restriction_{T_{g_n(y)} D_q} \right)} = \lim_{n \to \infty} \frac{\det \left(D_y f^{-n} \restriction_{T_y D_p} \right)}{\det \left(D_{h(y)} f^{-n} \restriction_{T_{h(y)} D_q} \right)}.$$

Let π^s be the projection over E^s parallel to E^u. By the invariance of the stable and unstable distributions Df^{-n} and π^s commute, i.e.,

$$\pi^s \circ D_y f^{-n} \cdot v = Df^{-n} \cdot \pi^s(v),$$

thus for any $y \in D_p$

$$D_y f^{-n} \restriction_{T_y D_p} = \left(\pi^s \restriction_{T_{f^{-n}(y)} f^{-n}(D_p)} \right)^{-1} \circ D_y f^{-n} \restriction_{E_y^s} \circ \pi^s \restriction_{T_y D_p}.$$

Then we can write

$$\det \left(D_y f^{-n} \restriction_{T_y D_p} \right) = \frac{\det \left(D_y f^{-n} \restriction_{E_y^s} \right) \det \left(\pi^s \restriction_{T_y D_p} \right)}{\det(\pi^s \restriction_{T_{f^{-n}(y)} f^{-n}(D_p)})}.$$

Now note that since $Tf^{-n}(D_p) \rightrightarrows E^s$, $n \to \infty$, we have

$$\det(\pi^s \restriction_{T_{f^{-n}(y)} f^{-n}(D_p)}) \rightrightarrows 1.$$

The same is true for every $y \in D_q$, thus we only need to show that uniformly we have

$$\lim_{n \to \infty} \frac{\det \left(D_y f^{-n} \restriction_{E^s} \right) \det \left(\pi^s \restriction_{T_y D_p} \right)}{\det \left(D_{g_n(y)} f^{-n} \restriction_{E^s} \right) \det \left(\pi^s \restriction_{T_{g_n(y)} D_q} \right)}$$

$$= \lim_{n \to \infty} \frac{\det(D_y f^{-n} \restriction_{E^s}) \det(\pi^s \restriction_{T_y D_p})}{\det(D_{h(y)} f^{-n} \restriction_{E^s}) \det(\pi^s \restriction_{T_{h(y)} D_q})}. \tag{4.10}$$

Observe that, since $g_n \rightrightarrows h$, $\det \left(\pi^s \restriction_{T_{g_n(y)} D_q} \right) \rightrightarrows \det \left(\pi^s \restriction_{T_{h(y)} D_q} \right)$ as $n \to \infty$. Therefore

$$\lim_{n \to \infty} \frac{\det(\pi^s \restriction_{T_y D_q})}{\det(\pi^s \restriction_{T_{g_n(y)} D_q})} = \frac{\det(\pi^s \restriction_{T_y D_q})}{\det(\pi^s \restriction_{T_{h(y)} D_q})}.$$

Then, to show (4.10) it is enough to prove that

$$\lim_{n\to\infty} \frac{\det\left(D_y f^{-n} \restriction_{E^s}\right)}{\det\left(D_{g_n(y)} f^{-n} \restriction_{E^s}\right)} = \lim_{n\to\infty} \frac{\det(D_y f^{-n} \restriction_{E^s})}{\det(D_{h(y)} f^{-n} \restriction_{E^s})}.$$

Applying (4.7) to y and $h(y)$ in the place of p and q, respectively, it follows that the uniform limit on the right side exists. Let us now show the same for the limit on the left side. Let us show that

$$\lim_{n\to\infty} \frac{\det\left(D_{h(y)} f^{-n} \restriction_{E^s}\right)}{\det\left(D_{g_n(y)} f^{-n} \restriction_{E^s}\right)} = 1, \qquad (4.11)$$

uniformly. By the chain rule, to show (4.11) is equivalent to prove

$$\lim_{n\to\infty} \prod_{k=0}^{n-1} \frac{\det(D_{f^{-k}(h(y))} f^{-k} \restriction_{E^s})}{\det(D_{f^{-k}(g_n(y))} f^{-n} \restriction_{E^s})} = 1,$$

which, in its turn, is equivalent to show

$$\lim_{n\to\infty} \sum_{k=0}^{n-1} \left| \det\left(D_{f^{-k}(h(y))} f^{-k} \restriction_{E^s}\right) - \det\left(D_{f^{-k}(g_n(y))} f^{-n} \restriction_{E^s}\right) \right| = 0 \qquad (4.12)$$

Now, as we have already seen

$$\left| \det\left(D_{f^{-k}(h(y))} f^{-k} \restriction_{E^s}\right) - \det\left(D_{f^{-k}(g_n(y))} f^{-n} \restriction_{E^s}\right) \right|$$
$$\leq D_2 d\left(f^{-k}(h(y)), f^{-k}(g_n(y))\right)^{\theta},$$

thus

$$\sum_{k=0}^{n-1} \left| \det\left(D_{f^{-k}(h(y))} f^{-k} \restriction_{E^s}\right) - \det\left(D_{f^{-k}(g_n(y))} f^{-n} \restriction_{E^s}\right) \right|$$
$$\leq D_2 \sum_{k=0}^{n-1} d\left(f^{-k}(h(y)), f^{-k}(g_n(y))\right)^{\theta}.$$

Now choose $\lambda < 1 < \lambda' < \mu' < \mu$. Since $f^{-n}(h(y)) \in W^u_{f^{-n}(y)}(\epsilon_n)$, $f^{-n}(g_n(y)) \in \mathcal{G}_{f^{-n}(y)}(\epsilon_n)$ and $\mathcal{G}$ is almost tangent to E^u we have $\epsilon_n < (\mu')^{-n}$ for n large enough. In particular, for large n we have on the one hand

$$d(f^{-n}(h(y)), f^{-n}(g_n(y))) \leq \epsilon_n < (\mu')^{-n}$$

On the other hand we have

$$d\left(f^{-k}(h(y)), f^{-k}(g_n(y))\right) = d\left(f^{n-k}(f^{-n}(h(y))), f^{n-k}(f^{-n}((g_n(y))))\right),$$

and for k large, $Tf^{-k}(D_q), \dots, Tf^{-n}(D_q)$ are close to E^s, implying that

$$d(f^{n-k}(f^{-n}(h(y))), f^{n-k}(f^{-n}((g_n(y))))) \le C \cdot (\lambda')^{n-k}(\mu')^{-n}$$

for some constant C. Finally, we conclude that

$$\sum_{k=0}^{n-1} \left| \det\left(D_{f^{-k}(h(y))} f^{-k} \upharpoonright_{E^s}\right) - \det\left(D_{f^{-k}(g_n(y))} f^{-n} \upharpoonright_{E^s}\right) \right|$$

$$\le D_2 \sum_{k=0}^{n-1} d\left(f^{-k}(h(y)), f^{-k}(g_n(y))\right)^\theta$$

$$\le D_2 C^\theta \sum_{k=0}^{n-1} \left((\lambda')^{\theta(n-k)}\right) \mu^{-n}$$

$$= D_3 \left((\lambda')^\theta + \dots + (\lambda')^{n\theta}\right) \mu^{-n\theta} = D_3(\lambda')^\theta \left(\frac{1-(\lambda')^{n\theta}}{1-(\lambda')^{n\theta}}\right) \mu^{-n\theta}.$$

As the right side of the next inequality is converging to zero when $n \to \infty$ we conclude that the series is indeed uniformly convergent, concluding the proof of the theorem. $\square$

4.1.4 The Ergodicity of Anosov Diffeomorphisms

Using the absolute continuity of stable and unstable foliations one can prove that every volume preserving $C^{1+\alpha}$-Anosov diffeomorphism is actually ergodic. The argument used to obtain ergodicity is usually called Hopf's argument.

Theorem 4.5 *Let $f : M \to M$ be a volume preserving $C^{1+\alpha}$ Anosov diffeomorphism of a smooth compact Riemannian manifold M. Then, f is ergodic.*

Proof Given a measurable function $\varphi : M \to \mathbb{R}$ denote by φ^+ and φ^- the forward and backward limit of the Birkhoff averages of φ, respectively, that is,

$$\varphi^+(x) := \lim_{n \to +\infty} \frac{1}{n} \sum_{k=0}^{n-1} \varphi\left(f^k(x)\right), \quad \varphi^-(x) = \lim_{n \to -\infty} \frac{1}{|n|} \sum_{k=0}^{n-1} \varphi\left(f^k(x)\right).$$

Also denote

$$\overline{\varphi}(x) = \lim_{n \to +\infty} \frac{1}{2n+1} \sum_{k=-n}^{n} \varphi\left(f^k(x)\right).$$

By Birkhoff ergodic theorem, the functions φ^+, φ^-, and $\overline{\varphi}$ are defined and are equal almost everywhere. Furthermore, to prove that f is ergodic it is enough to prove that given any continuous function $\varphi : M \to \mathbb{R}$, the function $\overline{\varphi}$ is constant almost everywhere. Let $\varphi : M \to \mathbb{R}$ be a continuous function and let M_0 be the full measure set where φ^+, φ^- and $\overline{\varphi}$ coincide.

Lemma 4.4 *If $x \in M_0$ and $z \in M$ is such that*

$$d\left(f^k(x), f^k(z)\right) \to 0, \quad k \to +\infty$$

then $\varphi^+(z) = \varphi^+(x)$. Consequently, if $x \in M_0$ and $z \in W^s(x)$, then $\varphi^+(z) = \varphi^+(x)$.

Proof Since $d(f^k(x), f^k(z)) \to 0$ as $k \to +\infty$, by the continuity of φ we have that $\lim_{n \to +\infty} |\varphi(f^k(x)) - \varphi(f^k(z))| = 0$. Therefore, the Cesàro averages of the later sequence also converge to the same limit, that is

$$\lim_{n \to +\infty} \frac{1}{n} \sum_{k=0}^{n-1} \left| \varphi\left(f^k(x)\right) - \varphi\left(f^k(z)\right)\right| = 0.$$

Thus

$$\left|\varphi^+(x) - \varphi^+(z)\right| = \left| \lim_{n \to +\infty} \frac{1}{n} \sum_{k=0}^{n-1} \varphi\left(f^k(x)\right) - \varphi\left(f^k(z)\right)\right|$$

$$\leq \lim_{n \to +\infty} \frac{1}{n} \sum_{k=0}^{n-1} \left|\varphi\left(f^k(x)\right) - \varphi\left(f^k(z)\right)\right|$$

$$= 0,$$

proving the statement of the lemma.   □

An analogous result is true for the unstable manifold.

Lemma 4.5 *If $x \in M_0$ and $z \in M$ is such that*

$$d\left(f^k(x), f^k(z)\right) \to 0, \quad k \to -\infty$$

then $\varphi^-(z) = \varphi^-(x)$. Consequently, if $x \in M_0$ and $z \in W^u(x)$, then $\varphi^-(z) = \varphi^-(x)$.

By the absolute continuity of the stable foliation (more precisely by Corollary 4.2), since $m(M_0) = 1$, for almost every point $x \in M$ and almost every point (with respect to the leaf measure) $z \in W^s(x)$ we have

$$m_{W^s}(x)(W^s(x) \setminus M_0) = 0. \qquad (4.13)$$

Now, for $r > 0$ small enough, we know that if $y \in B(x, r)$ then $W^u(y)$ is transversal to $W^s(x)$ and, consequently we can identify the transversal space $B(x, r)/W^u$ with $W^s(x)$. Therefore, using the absolute continuity of the unstable foliation and (4.13) we have

$$m \left(\bigcup_{y \in B(x,r), y \in W^s(x) \setminus M_0} W^u(y) \right) = 0.$$

In particular, for almost every point $z \in B(x, r) \cap M_0$ we have $W^u(z) \cap W^s(x) = \{y_0\}$ with $y_0 \in M_0$. By Lemmas 4.4 and 4.5 we have

$$\overline{\varphi}(z) \overset{z \in M_0}{=} \varphi^-(z) \overset{\text{Lemma 4.5}}{=} \varphi^-(y_0) \overset{y_0 \in M_0}{=} \varphi^+(y_0) \overset{\text{Lemma 4.4}}{=} \varphi^+(x) \overset{x \in M_0}{=} \overline{\varphi}(x),$$

for almost every $z \in B(x, y) \cap M_0$. That is, $\overline{\varphi}$ is constant almost everywhere in a neighborhood of x. Since M can be covered by a finite number of such neighborhoods where $\overline{\varphi}$ is constant almost everywhere, we conclude that $\overline{\varphi}$ is indeed constant almost everywhere in M and, therefore, f is ergodic as we wanted to show.

$\square$

4.2 The Kolmogorov Property for Anosov Diffeomorphisms

4.2.1 Leaf-Subordinated Partition and the Kolmogorov Property

Let $f : M \to M$ be a volume preserving $C^{1+\alpha}$ Anosov diffeomorphism of a compact Riemannian manifold M. In this section we will show that the pinsker partition of f is closely related to the measurable hull of the partition given by global stable/unstable manifolds of f.

Definition 4.9 If $f : M \to M$ is an Anosov diffeomorphism with stable (resp. unstable) foliation $\mathscr{F}^s$ (resp. $\mathscr{F}^u$), we denote by $\mathscr{H}(\mathscr{F}^s)$ (resp. $\mathscr{H}(\mathscr{F}^u)$) the measurable hull (see Definition 2.11) of the partition of M by global stable manifolds $\mathscr{F}^s(x)$, (resp. global unstable manifolds $\mathscr{F}^u(x)$) $x \in M$.

Definition 4.10 A leaf-subordinated partition associated with the global stable manifold is a measurable partition ξ such that

1. for μ-almost every $x \in M$, the element $\xi(x)$ in ξ which contains x is an open subset of $\mathscr{F}^s(x)$;
2. the partition $f\xi$ refines ξ, that is, $f\xi > \xi$;
3. $\bigvee_{i=0}^{\infty} f\xi = \varepsilon$;
4. $\Pi(\xi) = \mathscr{H}(\mathscr{F}^s)$.

To prove the existence of measurable partitions subordinated to the stable partition of an Anosov diffeomorphism we will need two results from measure theory which we quickly state below.

Theorem 4.6 (Borel–Cantelli) *Let $(X, \mathscr{A}, \mu)$ be a measure space and $\{A_n\}_{n \in \mathbb{N}}$ be a sequence of measurable subsets of X with*

$$\sum_{n=0}^{\infty} \mu(A_n) < \infty.$$

Then

$$\mu(\limsup_{n \to +\infty} A_n) = 0.$$

Lemma 4.6 (See [3, Lemma 9.4.2]) *Let $r > 0$ and let μ be a finite nonnegative Borel measure on $[0, r]$. For $0 < a < 1$, the Lebesgue measure of the set*

$$L_a := \left\{ s : 0 \leq s \leq r, \sum_{k=0}^{\infty} \mu\left([r - a^k, r + a^k]\right) < \infty \right\}$$

is exactly r.

Proof Consider the set

$$N_{a,k} := \left\{ s : 0 \leq s \leq r, \quad \mu\left(\left[r - a^k, r + a^k\right]\right) > \frac{\mu([0, r])}{k^2} \right\}.$$

Let us call an interval of length $2a^k$ centered at a point in $N_{a,k}$ a "bad interval." Observe that $N_{a,k}$ can be covered by bad intervals $J_{i,k}$, $1 \leq i \leq S(k)$ in a way that any point belongs to at most two bad intervals. Thus,

$$S(k) \cdot \frac{\mu([0, r])}{k^2} \leq \sum_{i=1}^{S(k)} \mu(J_{i,k}) \leq 2\mu([0, r])$$

and $m(N_{a,k}) \leq 2S(k)a^k$. Therefore

$$m(N_{a,k}) \leq 4k^2a^k \Rightarrow \sum_{k=0}^{\infty} m(N_{a,k}) < \infty.$$

By the Borel–Cantelli Theorem (Theorem 4.6), Lebesgue almost every point in L_a belong to at most a finite number of sets $N_{a,k}$ and, for such points, the series $\sum_{k=0}^{\infty} \mu([r - a^k, r + a^k])$ converges. □

Theorem 4.7 *Let $f : M \rightarrow M$ be a volume preserving $C^{1+\alpha}$ Anosov diffeomorphism of a compact Riemannian manifold. Then there exists a measurable partition η of M which is leaf subordinated to the global stable manifolds of f and $\Pi(\eta) > \pi(f)$.*

Proof Let $x \in M$ and $0 < r < l$ where l is a uniform lower bound on the sizes of the local stable manifolds. Now denote $B_0 = \bigcup_{w \in B(x,r)} W^s(w)$ and take

$$B := \bigcup_{n \geq 0} f^n(B_0).$$

By Theorem 4.5 f is ergodic, thus as $f(B) \subset B$ it follows that the set B must have either full or zero measure. Since $m(B_0) > 0$ we must have $m(B) = 1$. Therefore it is enough to construct a partition η of B.

Let $\widetilde{\xi}$ be the partition of B_0 by local stable manifolds $W^s(w)$, $w \in B(x,r)$ and let $\xi := \widetilde{\xi} \cup \{B \setminus B_0\}$ the partition of B obtained by adding the set $B \setminus B_0$ to the partition $\widetilde{\xi}$. We define the partition η by

$$\eta := \bigvee_{i=0}^{-\infty} f^i \xi$$

and we will show that η is a measurable partition subordinated to the stable manifolds of f.

Claim 1 For μ-almost every $w \in M$, the element $\xi(w)$ in ξ which contains w is an open subset of $\mathscr{F}^s(w)$.

First of all observe that since $m\left(\left[\bigcap_{j \geq 0} f^{-j}(B)\right] \setminus \left[\bigcup_{j \geq 0} f^{-j}(B_0)\right]\right) = 0$ then for m-almost every point $w \in B(x, r)$ we have

$$\eta(w) = f^n(W^s(f^{-n}(w))) \subset W^s(w)$$

for some $n \geq 0$. Now, consider the function $\beta(w)$ given by

$$\beta(w) = \inf_{n \geq 0} \left\{ r, \frac{\lambda^{-n}}{2} d(f^n(w), \partial B_0) \right\}.$$

Let $w \in B$ and $y \in W^s(w)$ with $d_s(y, w) \leq \beta(w)$. Then, for every $n \geq 0$ we have

$$d(f^n(y), f^n(w)) \leq d_s(f^n(y), f^n(w)) \leq \lambda^n \cdot d_s(y, w) \leq \frac{1}{2} d(f^n(w), \partial B_0).$$

This implies that $f^n(y) \in B_0$ if, and only if, $f^n(w) \in B_0$. In particular,

$$\xi(f^n(y)) = \xi(f^n(w)), \quad \forall n \geq 0.$$

It follows that if $d_s(y, w) \leq \beta(w)$ then, by the definition of η, we have

$$\eta(w) = \bigcap_{i \geq 0} f^{-i}(\xi(f^i(w))) = \bigcap_{i \geq 0} f^{-i}(\xi(f^i(y))) = \eta(y).$$

To conclude the proof of the first claim we just need to show that $\beta(w)$ is positive almost everywhere for a good choice of the initial r.

Consider μ the nonnegative measure on $[0, \beta]$ defined by

$$\mu(A) = m(\{y \in M : d(x, y) \in A\}).$$

Applying the Lemma 4.6 to $a := \lambda$ and to μ we have that $m(L_\lambda) = \beta$ where

$$L_\lambda := \left\{ s : 0 \leq s \leq r, \sum_{k=0}^{\infty} m \left(\{y \in M : |d(x, y) - s| < \lambda^k\} \right) < \infty \right\}.$$

Since m is f-invariant we have $m(\{y \in M : |d(x, y) - s| < \lambda^k\}) = m(\{y \in M : |d(x, f^k(y)) - s| < \lambda^k\})$ thus

$$L_\lambda := \left\{ s : 0 \leq s \leq r, \sum_{k=0}^{\infty} m \left(\{y \in M : |d(x, f^k(y)) - s| < \lambda^k\} \right) < \infty \right\}.$$

Let $D > 0$ be such that

$$d(w, \partial B_0) < \tau \Rightarrow |d(x, w) - r| < D\tau, \quad 0 < \tau \leq r < \beta.$$

Then

$$\widetilde{L}_\lambda := \left\{ s : 0 \leq s \leq r, \sum_{k=0}^{\infty} m \left(\{y \in M : d(f^k(y), \partial B_0) < D^{-1} \cdot \lambda^k\} \right) < \infty \right\} \subset L_\lambda.$$

Now observe that, for any $s \in \widetilde{L}_\lambda$, applying the Borel–Cantelli theorem (see Theorem 4.6) to the sequence

$$A_k := \left\{ y \in M : d(f^k(y), \partial B_0) < D^{-1} \cdot \lambda^k \right\}$$

we have $m(\limsup_{k \to +\infty} A_k) = 0$, that is, for almost every $y \in M$ there exists only a finite number of natural numbers with

$$d\left(f^k(y), \partial B_0 \right) < D^{-1} \cdot \lambda^k$$

and consequently $\beta(y)$ is positive. Therefore, we just need to take $r \in \widetilde{L}_\lambda$ to obtain that $\beta(w)$ is positive for almost every w. In particular, each set $\eta(w)$ contains an open subset of $W^s(w)$.

Claim 2 $f\eta > \eta$. Indeed, by the definition of η we have $f\eta = f\left(\bigvee_{i=0-\infty} f^i \xi \right) = f\xi \vee \eta > \eta$, which proves the second claim.

Claim 3 $\bigvee_{i=0}^{\infty} f^i \eta = \varepsilon$. Observe that it is enough to prove that for almost every pair of points $y_1, y_2 \in B_0$ there exists an integer n with $z_1 \notin f^n \eta(z_2)$. By poincaré's Recurrence Theorem we know that almost every point $z \in B_0$ returns infinitely often to B_0. If $n_1, n_2, \dots$ are the returning times, then $f^{n_i} \xi(z) \subset f^{n_i}(\xi(f^{-n_i}(z)) \subset f^{n_i}(W^s(f^{-n_i}(z)) \cap B_0)$, so that

$$\operatorname{diam}(f^{n_i} \eta(z)) \to 0, \quad i \to \infty.$$

In particular, given any two points z_1, z_2 in the full measure set given by the Poincaré's Recurrence Theorem as done above, for some n big enough the diameter of $f^n \eta(z_1)$ is smaller than the distance between z_1 and z_2 and, consequently, $z_2 \notin f^n \eta(z_2)$ as we wanted to show.

Claim 4 $\Pi(\eta) = \mathscr{H}(\mathscr{F}^s) > \pi(f)$.

First of all observe that $\eta^- = \eta$, thus $\bigwedge_{i > -\infty}^{0} f^i \eta^- = \bigwedge_{i > -\infty}^{0} f^i \eta$. Now set $\widetilde{\eta} = \bigwedge_{i > -\infty}^{0} f^i \eta$. Thus we have $\widetilde{\eta}|B_0 < \widetilde{\xi}$ and, by the definition of the global manifold $\mathscr{F}^s(z)$, one can see that for almost every $z \in B$ we have

$$\mathscr{F}^s(z) \subset \widetilde{\eta}(z).$$

Thus $\mathscr{F}^s|B > \widetilde{\eta}$ which implies $\mathscr{H}(\mathscr{F}^s|B) > \mathscr{H}(\widetilde{\eta}) = \Pi(\eta)$.

But from the measurability of $\widetilde{\eta}$ we obtain $\mathscr{H}(\mathscr{F}^s|B) > \widetilde{\eta}$. On the other hand, we have proved in the first item that $\eta > \mathscr{F}^s|B$. Therefore, $\widetilde{\eta} > \mathscr{F}^s|B$ which implies $\Pi(\eta) = \mathscr{H}(\widetilde{\eta}) > \mathscr{H}(\mathscr{F}^s|B)$. That is, in terms of equality modulo zero we have

$$\mathscr{H}(\mathscr{F}^s) = \Pi(\eta).$$

Finally, we have already proved in claims 2 and 3 that η is an exhaustive partition. Now, from Theorem 2.15 we have $\bigwedge_{i > -\infty}^{0} f^i \eta > \pi(f)$ which implies $\Pi(\eta) > \pi(f)$.

$\square$

Theorem 4.8 *Let $f : M \to M$ be a $C^{1+\alpha}$ volume preserving Anosov diffeomorphism of a compact Riemannian manifold M. Then f is a Kolmogorov automorphism.*

Proof Let $x \in M$ be a certain fixed point in M and take $B_0^s := \bigcup_{w \in B(x,r)} W^s(w)$ and $B_0^u := \bigcup_{w \in B(x,r)} W^u(w)$, where $r > 0$ is small enough according to the construction of the subordinated partition made in Theorem 4.7. Let

$$P := B_0^s \cup B_0^u$$

and let η_s be the partition subordinated to the stable foliation of f constructed as in Theorem 4.7 and analogously η_u the partition subordinated to the stable foliation of f^{-1} (consequently it is a partition subordinated to the unstable foliation of f). By Theorem 4.7 we have

$$\pi|B_0^s < \eta_s, \quad \pi|B_0^u < \eta_u.$$

In particular we have

$$\pi|P < \eta_s \wedge \eta_u. \tag{4.14}$$

Now, consider an element of $\eta_s(z) \in \eta_s$ for which the conditional measure on the stable manifold is absolutely continuous with respect to the leaf measure. Observe that since

$$\eta_s \wedge \eta_u(z) = \bigcup_{w \in \eta_s(z)} \eta_u(w)$$

and since $\eta_u(w)$ is an open set in $\mathscr{F}^u(w)$ we have that the volume measure of a typical element of $\eta_s \wedge \eta_u$ is positive. By (4.14) we conclude that the pinsker partition π has an element of positive volume measure. Since π is invariant, all the elements must have the same measure and, consequently, π has a finite number of elements of positive measure which are cyclicly permuted by f. In particular, there exist an integer n and an element $\varXi \in \pi$ such that

$$m(\varXi) > 0 \quad \text{and} \quad f^n(\varXi) = \varXi.$$

But since f is an Anosov diffeomorphism, f^n is also a $C^{1+\alpha}$ volume preserving Anosov diffeomorphism and Theorem 4.5 implies that f^n is ergodic and, consequently, $m(\varXi) = 1$. Therefore π is the trivial partition and f is Kolmogorov as we wanted to show.

□

4.3 Volume Preserving Anosov Diffeomorphisms Are Bernoulli

We now proceed to the proof of the main result of this chapter, namely that every $C^{1+\alpha}$ volume preserving Anosov diffeomorphism is Bernoulli.

Theorem 4.9 *Let M be a compact Riemannian manifold and $f : M \to M$ be a volume preserving $C^{1+\alpha}$ Anosov diffeomorphism. Then f is a Bernoulli automorphism.*

The proof of this fact is parallel to the proof of Bernoulli property for linear ergodic hyperbolic automorphisms of $\mathbb{T}^2$ made in Chap. 2. Indeed, the reader should notice that the stable and unstable bundles are a generalization of the eigenspaces of a hyperbolic automorphism and the stable and unstable foliations are a generalization of the stable and unstable manifolds of a hyperbolic automorphism. Thus the path to show Bernoulli property for an Anosov diffeomorphism follows the same philosophy of that for the hyperbolic automorphism, in particular most of the lemmas necessary to prove Bernoullicity of ergodic hyperbolic automorphisms are extended to the Anosov setting by taking the stable and unstable directions in the role of the eigenspaces of the hyperbolic automorphism.

In the light of these observations it is crucial to note that, for the ergodic hyperbolic automorphism setting, there is one step of the proof which heavily depends on the application of Fubbini's Theorem to the unstable manifolds of the automorphism (see Lemma 5.8). As we have mentioned before, Fubbini's Theorem is not true for every continuous foliation of a given manifold, therefore to generalize this particular step to the Anosov context we must rely upon the absolute continuity property of the stable and unstable foliations of $C^{1+\alpha}$ Anosov diffeomorphisms (Theorem 4.4). As we will see in the last chapter, the absence of absolute continuity for certain invariant foliations is one of the obstructions for the Bernoulli property to occur in more general contexts.

Let f be as in the statement of Theorem 4.9. In what follows the constants $\lambda < 1 < \mu$ are the constants related to the $C^{1+\alpha}$ Anosov diffeomorphism f according to the statement of Theorem 4.1.

4.3.1 Partition by Rectangles

In order to prove Bernoullicity of an Anosov diffeomorphism we will work with partitions which are suitable in this context, more precisely partitions by sets with local product structure. Sets of this type will be called rectangles.

Definition 4.11 A measurable set $R \subset M$ is called a δ-rectangle at a point $w \in M$ if it satisfies the following conditions:

(i) $w \in R \subset B(x, \delta)$;
(ii) for any $x, y \in R$ we have

$$W^s(x) \cap W^u(y) = \{z\} \subset R.$$

From now on, for the sake of simplicity, we will make the following convention: If R is a rectangle and $x \in R$, then $W^\tau(x) \cap R$ denotes the connected component of $W^\tau(x) \cap R$ which contains the point x, $\tau \in \{s, u\}$.

In what follows we show that M admits a partition by δ-rectangles for any $\delta > 0$. The proof we present here is the same proof given in Lemmas 9.5.5, 9.5.6, and 9.5.7 of [3] but restricted to the setting of Anosov diffeomorphisms.

Lemma 4.7 *For any sufficiently small $\delta > 0$, one can find disjoint rectangles $R_1, R_2, \ldots, R_m$ such that $\{R_1, R_2, \ldots, R_m\}$ is a partition of M.*

Proof Let $w \in M$ an arbitrary point. By item (3) of Remark 4.1 for a sufficiently small $\delta_w > 0$ there exists $r_w = r(\delta_w) > 0$ such that for any $y, z \in B(w, r_w)$ we have

$$W^s(y) \cap W^u(z) \in B(w, \delta_w).$$

Now take

$$R(w) = \{x \in M : \text{there are } y, z \in B(w, r) \text{ with } x = W^s(y) \cap W^u(z)\}.$$

We claim that $R(w)$ is a rectangle centered at w. Obviously $w \in R(w)$ since $w = W^s(w) \cap W^u(w)$. Now let $x_1, x_2 \in R(w)$. By the definition of $R(w)$ there are points $y_1, z_1, y_2, z_2 \in B(w, \delta_w)$ such that

$$x_1 = W^s(y_1) \cap W^u(z_1), \quad x_2 = W^s(y_2) \cap W^u(z_2).$$

Now for δ_w small enough we have

$$W^s(x_1) \cap W^u(x_2) = W^s(y_1) \cap W^u(z_2) \in R(w),$$

concluding that $R(w)$ is indeed a rectangle centered at w.

Given $\delta > 0$ small enough, consider the covering of M by the family of open balls $\{B(w, \delta_w) : w \in M, \delta_w < \delta\}$. Since M is compact M is covered by a finite number of such open balls and, consequently, by a finite number of rectangles of the family $\{R(w) : w \in M\}$ constructed above. More precisely, we can take $w_1, \ldots, w_m$ such that

$$M = \bigcup_{i=1}^{m} B(w_i, \delta_{w_i}) = \bigcup_{i=1}^{m} R(w_i). \tag{4.15}$$

Now, in order to obtain a partition by rectangles we need to analyze what happens when two rectangles intersect. Let $R = R(w_i)$ and $R' = R(w_j)$, $i \neq j$, and assume that $R \cap R' \neq \emptyset$. Consider the set

$$\widetilde{R} = \bigcup_{y \in R \cap R'} W^u(y).$$

The sets

$$R_1 = R \cap R', \quad R_2 = R \setminus \widetilde{R}, \quad R_3 = R' \setminus \widetilde{R}, \quad R_4 = (\widetilde{R} \cap R) \setminus (R' \cap R)$$

and

$$R_5 = (\widetilde{R} \cap R') \setminus (R' \cap R)$$

are disjoint and their union is $R \cup R'$. It is not difficult to prove that R_i, $1 \leq i \leq 5$, are rectangles and we leave this fact as an exercise to the reader. Therefore we may replace any two rectangles $R(w_i)$ and $R(w_j)$ in (4.15) which are not disjoint by five disjoint rectangle. This yields a partition of M by δ-rectangles as we wanted. $\quad\square$

4.3.2 Intersections Along the Unstable Direction

In the previous subsection we showed that given any $\delta > 0$ arbitrarily small one may take a finite partition, say β, of M by δ-rectangles. Here we will take an arbitrary finite partition α by measurable sets whose boundaries are piecewise smooth and we want to see what happens when we iterate and refine this partition. Intuitively speaking, as we iterate an element A of α the set A stretches in the unstable direction and, at some point, it should cross most of the rectangles in β along the unstable direction. This is what we show in Lemma 4.8 below.

Definition 4.12 Let $\alpha = \{A_1, \ldots, A_k\}$ be a partition of M. We say that α is an SB-partition (smooth boundary partition) if for each A_i, $1 \leq i \leq k$, the boundary of A_i is piecewise smooth.

Definition 4.13 Given a rectangle R and a subset $E \subset M$, we say that the intersection $E \cap R$ is a u-tubular intersection if for every $x \in E \cap R$ we have

$$W^u(x) \cap R \subset R \cap E.$$

Lemma 4.8 *Let α be an SB-partition, R a rectangle, and $\delta > 0$ a given real number. There exists a sufficiently large natural number N_1 such that for any $N' > N > N_1$ and δ-almost every atom $P \in \bigvee_N^{N'} f^k \alpha$ there exists a subset $E \subset P$ with*

(i)

$$\frac{m(E)}{m(P)} \geq 1 - \delta;$$

(ii) the intersection $E \cap R$ is a u-tubular intersection.

Proof Observe that by the invariance of the stable and unstable leaves the u-tubularity of an intersection is not affected by applying f. For $P \in \alpha$ denote by G_k^P the non u-tubular intersections of R and $f^k(P)$, more specifically

$$G_k^P = \{y \in R \cap f^k(P) : W^u(y) \cap R \not\subset R \cap f^k(P)\}.$$

We now estimate the measure of G_k.

Claim For a certain constant $C > 0$ and a certain $\mu' > 1$, given any $y \in f^{-k}(G_k^P)$ we have

$$d(y, \partial P) \leq C \cdot (\mu')^{-k}. \tag{4.16}$$

Proof Consider $y \in f^{-k}(G_k^P)$ arbitrary. Since y is in the non-tubular intersection set we have that $W^u(y) \cap f^{-k}(R) \not\subset f^{-k}(R) \cap P$. Since the connected component $W^u(y) \cap f^{-k}(R)$ contains a point in P and a point w outside P then $W^u(y) \cap f^{-k}(R)$ intersects the boundary of P in a point z, in particular

$$d(y, w) \leq d(y, z) + d(z, w).$$

Now, take $\varepsilon < \mu - 1$ and define $\mu' := \mu - \varepsilon$ and $D := C(\varepsilon)$ where $C(\varepsilon)$ is given by Theorem 4.1. Then, by the third item of Theorem 4.1 we have

$$d(y, w) \leq D \cdot (\mu')^{-k} d\left(f^k(y), f^k(w)\right) \leq D \cdot \mathrm{diam}(R) \cdot (\mu')^{-k}.$$

Thus it is enough to take $C = D \cdot \mathrm{diam}(R)$.   △

Now the rest of the proof follows ipsis literis the same lines as in the proof of Lemma 3.3.   □

4.3.3 Construction of the Function θ and Conclusion of the Proof

To construct the function θ with the desired properties we will first prove an analogue of Lemma 3.4 of Chap. 2. The reader must observe that the crucial difference between the next lemma and Lemma 3.4 is that we cannot use the classical Fubbini's Theorem in the Anosov context, instead we rely upon the absolute continuity property of the stable and unstable foliations. As a consequence the function θ constructed here is not necessarily measure preserving, as obtained in Lemma 3.4, but instead it has a controlled Jacobian.

Lemma 4.9 *Given $\delta_1 > 0$, there exists $\delta_2 > 0$ such that if R is a δ_2-rectangle and E is a u-tubular subset of R, then there exists an injective map $\theta : E \to R$ such that*

(1) $|\operatorname{Jac}(\theta)(x) - 1| \leq \delta_1$ for any $x \in E$;
(2) for every $k \geq 1$ and $x \in E$

$$d\left(f^k(\theta x), f^k(x)\right) < \delta_1.$$

Proof Choose $w_0 \in R$ and consider the holonomy map

$$\pi^u_{w,w_0} : X \to W^s(w_0)$$

where $X \subset W^s(w)$ is a measurable set (see Definition 4.5). By Theorem 4.3 applied to the family of local unstable manifolds we may take $\delta_2 > 0$ small enough so that

$$|\operatorname{Jac}(\pi^u_{w,w_0})(x) - 1| \leq \frac{1}{3}\delta_1, \quad \forall x \in X. \tag{4.17}$$

If δ_2 is sufficiently small then, by the third item of Theorem 4.1, we clearly have

$$d\left(f^k(y), f^k(x)\right) < \delta_1,$$

for all $x \in E$ and $y \in W^s(x) \cap R$. Thus for the second item to be satisfied it is enough to take δ_2 small enough and guarantee that $\theta(x) \in \mathscr{F}^s(x)$.

Let E be a measurable subset of positive measure, which intersects the rectangle R in a u-tubular subset. Since the intersection $E \cap W^s(w_0)$ has positive $m_{W^s}(w_0)$-measure we can take a bijective map

$$\theta_0 : E \cap W^s(w_0) \to W^s(w_0) \cap R$$

which preserves the normalizes measures, that is,

$$(\theta_0)_* m_{W^s}(w_0)(\cdot|E) = m_{W^s}(w_0)(\cdot|R).$$

where $m_{W^s}(w_0)(\cdot|E) = \frac{m_{W^s}(w_0)}{m_{W^s}(w_0)(E \cap W^s(w_0))}$ and $m_{W^s}(w_0)(\cdot|R) = \frac{m_{W^s}(w_0)}{m_{W^s}(w_0)(R \cap W^s(w_0))}$.

Now, for every $y \in E$ let,

$$z(y) = W^u(y) \cap W^s(w_0) \in R \cap W^s(w_0).$$

Then $z(y) \in E \cap W^s(w_0)$ and we define

$$\theta(y) := W^u(\theta_0(z(y))) \cap W^s(y) = \pi^u_{w_0, y} \circ \theta_0 \circ \left(\pi^u_{w_0, y}\right)^{-1}(y). \qquad (4.18)$$

Now, from (4.17) and (4.18) it follows that

$$|\operatorname{Jac}(\theta)(y) - 1| \le \delta_1$$

as we wanted to show.

$\square$

Lemma 4.10 *Let $\varepsilon, \varepsilon' > 0$ and α be an SB-partition. Then, there exists N such that for all $N' \ge N$ and for ε-almost every atom $A \in \bigvee_N^{N'} f^k \alpha$ there exists a subset $E \subset A$ and an injective function $\theta : E \to M$ such that*

(1) θ *is $c\varepsilon$- measure preserving for a certain constant $c > 0$;*
(2)

$$\frac{m(E)}{m(A)} > 1 - \varepsilon;$$

(3) for every $k \ge 0$ and $x \in E$,

$$d\left(f^k(\theta x), f^k(x)\right) < \varepsilon'.$$

Proof The proof is very much similar to the proof of Lemma 3.5. The differences are: here we take the partition β to be a partition by rectangles and the functions θ_i are almost measure preserving, while in the proof of Lemma 3.5 they could be taken to be measure preserving. For the sake of completeness we repeat the whole argument in detail here.

Given $\varepsilon' > 0$, by applying Lemma 4.9 for $\delta_1 := \varepsilon'$ we obtain $\delta_2 > 0$ such that if R is a rectangle of diameter smaller than δ_2 and E is a u-tubular subset of R, then there exists an injective map $\theta : E \to R$ satisfying the properties (1) and (2) of Lemma 4.9.

Take a partition $\beta := \{R_1, \ldots, R_b\}$ where $R_1, \ldots, R_b$ are rectangles with diameter smaller than or equal δ_2. Consider

$$\gamma := \varepsilon \cdot b^{-1} \cdot \min\{m(R_i) : 1 \le i \le b\}.$$

For each $1 \leq i \leq b$, let $N_1^i > 0$ be provided by Lemma 4.8 applied for $R := R_i$ and $\delta := \gamma$. Thus, taking $N_1 = \max\{N_1^i : 1 \leq i \leq b\}$ we have that, given $N' > N \geq N_1$, for ε-almost every atom P of $\bigvee_N^{N'} f^k \alpha$ and for each $1 \leq i \leq b$ there exists a subset $E_i \subset P$, constructed in the proof of Lemma 4.8, such that

$$\frac{m(E_i)}{m(P)} > 1 - \gamma \tag{4.19}$$

and $E_i \cap R_i$ is u-tubular in R_i for all $1 \leq i \leq b$. For such an atom P fixed let

$$E := \bigcup_{i=1}^{b} (E_i \cap R_i) \subset P. \tag{4.20}$$

It is easy to see that E intersects each R_i in a u-tubular subset. Furthermore, observe that by the construction of E_i (see Lemma 4.8) for $j \neq i$ we have $P \cap R_j = E_i \cap R_j$ since every point in $R_j \cap P$ is not in the non-tubular intersection with respect to R_i, thus we have by (4.19)

$$\frac{m(E)}{m(P)} = \sum_{i=1}^{b} \frac{m(E_i \cap R_i)}{m(P)} = \sum_{i=1}^{b} \frac{m(E_i) - \sum_{j \neq i} m(P \cap R_j)}{m(P)}$$

$$= \sum_{i=1}^{b} \frac{m(E_i)}{m(P)} - \sum_{i=1}^{b} \sum_{j \neq i} \frac{m(P \cap R_j)}{m(P)} \tag{4.21}$$

$$> b \cdot (1 - \gamma) - (b - 1) \sum_{j=1}^{b} \frac{m(P \cap R_j)}{m(P)}$$

$$> 1 - b\gamma > 1 - \varepsilon \cdot \min\{m(R_i) : 1 \leq i \leq b\}.$$

Also, for such $E \subset P$ we have

$$\left| \frac{m(E \cap R_i)}{m(E)} - m(R_i) \right| = \left| \left[\frac{m(P \cap R_i)}{m(P)} - m(R_i) \right] \frac{m(P)}{m(E)} + m(R_i) \left(\frac{m(P)}{m(E)} - 1 \right) \right|$$

$$< \frac{1}{1 - \varepsilon} \cdot \left| \frac{m(P \cap R_i)}{m(P)} - m(R_i) \right| + \frac{1}{1 - \varepsilon} - 1 \tag{4.22}$$

Now, by Theorem 3.4 f is Kolmogorov and, consequently, by Lemma 2.1 it follows that every finite partition, in particular the partition α, is Kolmogorov. That is, for each $1 \leq i \leq b$, given $\xi_i > 0$ there exists $N_0^i > 0$ such that for any $N' > N \geq N_0^i$ and ξ_i-almost every element $P \in \bigvee_{k=N}^{N'} f^k \alpha$ we have

$$\left| \frac{m(P \cap R_i)}{m(P)} - m(R_i) \right| \leq \xi_i. \tag{4.23}$$

Let $D > 0$ be a constant such that

$$D(1 - \varepsilon) \cdot \min\{m(R_i) : 1 \leq i \leq b\} > 1$$

and take

$$\xi_i < \varepsilon \cdot (D(1 - \varepsilon) \cdot \min\{m(R_i) : 1 \leq i \leq b\} - 1), \quad 1 \leq i \leq b,$$

and $N_2 := \max\{N_1, N_0^1, \ldots, N_0^b\}$. For ξ_i small enough, by (4.20)–(4.23) we have that for any $N' > N \geq N_2$ and ε-almost every element $P \in \bigvee_{k=N}^{N'} f^k \alpha$ there exists a subset $E \subset P$ intersecting each R_i in a u-tubular way satisfying

$$\frac{m(E)}{m(P)} > 1 - \varepsilon \cdot \min\{m(R_i) : 1 \leq i \leq b\} \geq 1 - \varepsilon \tag{4.24}$$

and, for each $1 \leq i \leq b$ we have

$$\left| \frac{m(E \cap R_i)}{m(E)} - m(R_i) \right| < \frac{1}{1 - \varepsilon} \cdot \xi_i + \frac{1}{1 - \varepsilon} - 1$$

$$< \frac{1}{1 - \varepsilon}(\varepsilon \cdot (D(1 - \varepsilon) \cdot \min\{m(R_i) : 1 \leq i \leq b\} - 1) + 1)$$

$$- 1 = D \cdot \varepsilon \cdot \min\{m(R_i) : 1 \leq i \leq b\} \leq D \cdot \varepsilon \cdot m(R_i).$$

In particular we have

$$\left| \frac{m(E)m(R_i)}{m(E_i \cap R_i)} - 1 \right| \leq \frac{D \cdot \varepsilon}{1 - D \cdot \varepsilon}, \quad 1 \leq i \leq b. \tag{4.25}$$

Let $\theta_i : E_i \cap R_i \to R_i$, $i = 1, \ldots, b$ be the injective maps constructed in Lemma 4.9. Define the map

$$\theta : E \to M, \theta(y) = \theta_i(y) \text{ if } y \in E_i.$$

By Lemma 4.9, for a measurable set $B \subset E$ we have:

$$\left| \frac{m(\theta_i(B_i))}{m(R_i)} \cdot \frac{m(E_i \cap R_i)}{m(B_i)} - 1 \right| \leq \varepsilon',$$

where $B_i := B \cap R_i$. Then,

$$\left| \frac{m(\theta_i(B_i))m(E)}{m(B_i)} - 1 \right| \leq \left| \frac{m(\theta_i(B_i))m(E_i \cap R_i)}{m(B_i)m(R_i)} - 1 \right|$$

$$+ \left| \frac{m(\theta_i(B_i))m(E_i \cap R_i)}{m(B_i)m(R_i)} \cdot \left(\frac{m(E)m(R_i)}{m(E_i \cap R_i)} - 1 \right) \right|$$

$$\leq \varepsilon' + (1 + \varepsilon') \left| \frac{m(E)m(R_i)}{m(E_i \cap R_i)} - 1 \right|.$$

Thus, by (4.25) we have

$$\left| \frac{m(\theta_i(B_i))m(E)}{m(B_i)} - 1 \right| \leq \varepsilon' + (1 + \varepsilon') \frac{D \cdot \varepsilon}{1 - D \cdot \varepsilon} < c \cdot \varepsilon \qquad (4.26)$$

for a certain constant $c > 0$. Finally, observe that

$$\left| \frac{m(\theta(B))m(E)}{m(B)} - 1 \right| = \left| \frac{\sum_{i=1}^{b} m(\theta_i(B_i))m(E)}{m(B)} - 1 \right|$$

$$= \left| \sum_{i=1}^{b} \left[\frac{m(\theta_i(B_i))m(E)}{m(B_i)} - 1 \right] \cdot \frac{m(B_i)}{m(B)} \right|.$$

Then, by (4.26) we have

$$\left| \frac{m(\theta(B))m(E)}{m(B)} - 1 \right| < \sum_{i=1}^{b} c \cdot \varepsilon \cdot \frac{m(B_i)}{m(B)} = c \cdot \varepsilon,$$

concluding that θ is indeed $c \cdot \varepsilon$-measure preserving for a certain constant $c > 0$. The third item is an immediate consequence of the construction of θ. $\qquad \square$

Lemma 4.11 *If α is an SB-partition and $\varepsilon > 0$, then there exists N such that for every $N' \geq N$ and ε-almost every atom $A \in \bigvee_N^{N'} f^k \alpha$ we have*

$$d(\{f^{-i}\alpha | A\}_1^n, \{f^{-i}\alpha\}_1^n) \leq \varepsilon$$

for every $n \geq 1$. In other words, any SB-partition is VWB for A.

Proof The proof of this result is identical to the proof of Lemma 3.6 but using the respective auxiliary lemmas in the context of Anosov diffeomorphisms.

$\qquad \square$

This concludes the proof of Theorem 4.9 as we wanted.

References

1. Anosov, D.V.: Ergodic properties of geodesic flows on closed Riemannian manifolds of negative curvature. Dokl. Akad. Nauk SSSR **151**, 1250–1252 (1963)
2. Anosov, D.V., Sinaĭ, J.G.: Certain smooth ergodic systems. Uspehi Mat. Nauk. **22**, 107–172 (1967)
3. Barreira, L., Yakov, P.: Nonuniform Hyperbolicity. Dynamics of Systems with Nonzero Lyapunov Exponents. Cambridge University Press, Cambridge (2007)
4. Katok, A., Hasselblatt, B.: Introduction to the Modern Theory of Dynamical Systems. Encyclopedia of Mathematics and its Applications, vol. 54. Cambridge University Press, Cambridge (1995)
5. Pugh, C., Shub, M.: Ergodicity of Anosov actions. Invent. Math. **15**, 1–23 (1972)

Chapter 5
State of the Art

Abstract This chapter is devoted to present some other results concerning the equivalence of the Kolmogorov and the Bernoulli property for systems which preserve a smooth measure and admit some level of hyperbolicity. We define the class of non-uniformly hyperbolic diffeomorphisms (resp. flows), the class of smooth maps (resp. flows) with singularities, and the class of partially hyperbolic diffeomorphisms derived from Anosov, and present the state of art of the problem inside each of this classes. In each case we briefly comment on the similarities with the Anosov case as well as the central difficulties that appear along the arguments. The class derived from Anosov diffeomorphisms is the one for which the results differ the most from the results for Anosov diffeomorphisms, therefore we go deeper in this particular case and prove the key results which allow us to overcome the absence of complete hyperbolicity along the center direction.

5.1 Non-uniformly Hyperbolic Diffeomorphisms

We have seen in Chap. 4 that Kolmogorov implies Bernoulli in the Anosov diffeomorphism case. We now want to show how the result also works for a much more general case, but which proofs the reader would find not too difficult if he or she has read the previous two chapter of this book. For the basic definitions and results in this subsection the reader may also take a look at the books Barreira and Pesin [4] and Katok and Hasselblatt [17].

As usual $f : M \to M$ is a diffeomorphism. If the limit

$$\lambda(x, v) := \lim_{n \to \infty} \frac{1}{n} \log ||Df^n(x) \cdot v||, v \neq 0$$

exists we call $\lambda(x, v)$ the Lyapunov exponent of f at x in the direction of v. Observe from the definition of this limit that the Lyapunov exponent in this direction tries to measure the exponential growth of the function in that directions. Recall that in the Anosov case if we consider v in the stable direction it turns out that $\lambda(x, v) < 0$. In

G. Ponce, R. Varão, *An Introduction to the Kolmogorov–Bernoulli Equivalence*,
SpringerBriefs in Mathematics, https://doi.org/10.1007/978-3-030-27390-3_5

fact if the Lyapunov exponent is non-zero it means that we observe, for sufficiently large n, some hyperbolicity but we cannot conclude anything for the zero Lyapunov exponent.

The next result, which is a consequence of a more general result known as the Oseledets Theorem, is quite surprising and states that at least in the measurable point of view the existence of Lyapunov exponents is a common thing. The first version of the Oseledets Theorem was proved in 1965 by Valery Oseledets (also spelled "Oseledec") and was later extended to other situations by several mathematicians.

Theorem 5.1 (Oseledets Theorem) *Let $f : M \to M$ be a C^1 diffeomorphism defined on a compact Riemannian manifold M. The set of points $x \in M$ which satisfies:*

- *there exists a splitting*

$$T_x M = E_1(x) \oplus \ldots \oplus E_{k(x)};$$

- $Df_x E_i(x) = E_i(f(x)), k(x) = k(f(x));$
- *there exist the limits*

$$\lambda_i(x) := \lim_{n \to \infty} \frac{1}{n} \log \|Df^n(x) \cdot v\| = \lim_{n \to -\infty} \frac{1}{n} \log \|Df^n(x) \cdot v\|,$$

for all $v \in E_i(x) \backslash \{0\}$,

is a set of full measure for any f-invariant probability measure.

In the mid of 1970 Yakov Pesin developed, in a series of landmark works [23–25], a general theory for systems which do not admit a uniformly hyperbolic splitting, as the Anosov systems have, but instead admit at least a hyperbolic like structure whose expansion and contraction constants may change along the orbit of a given point at a certain rate. This theory, which is nowadays called Pesin's Theory , allows one to translate the asymptotic characteristics of f given, for example, by the Lyapunov exponents of f into dynamical characteristics of f. One of the results reflecting this idea is the stable manifold theorem proved by Pesin which establishes the existence of "contracting" embedded disks tangent to a certain subspace associated with the non-zero Lyapunov exponents of a certain point.

Theorem 5.2 (Stable Manifold Theorem [4]) *Let $f : M \to M$ be a $C^{1+\alpha}$ diffeomorphism on a compact manifold M, then there exists an embedded disk $W_{loc}^s(x)$ through x and there exist $C_x > 0$ such that*

- $W_{loc}^s(x)$ *is tangent to $E_x^2 := \oplus_{\lambda_i(x) < 0} E_x^i$;*
- $d(f^n(y), f^n(z)) \le C_x e^{-n\tau_x} d(y, z)$ *for all $y, z \in W_{loc}^s(x)$;*
- $f(W_{loc}^s(x)) \subset W_{loc}^s(f(x));$
- *if all Lyapunov exponents are non-zero, then*

$$W^s(x) = \bigcup_{n=0}^{\infty} f^{-n}(W^u_{loc}(f^n(x)),$$

where $W^s(x) = \{y \in M | d(f^n(x), f^n(y)) \to 0, \ as \ n \to \infty\}$.

We call $W^s_{loc}(x)$ *local Pesin stable manifold* and we define the *local Pesin unstable manifold* $W^u_{loc}(x)$ as the Pesin stable manifold for f^{-1}.

Definition 5.1 Let $f : M \to M$ be a $C^{1+\alpha}$ diffeomorphism on a compact Riemannian manifold M. The diffeomorphism f is said to be non-uniformly hyperbolic on an f-invariant Borel subset $Y \subset M$ if there exists

(a) Borel functions $\lambda, \mu, C, K : Y \to (0, \infty)$ and $\varepsilon : Y \to [0, \varepsilon_0]$ with $\varepsilon_0 > 0$ and
(b) subspaces $E^s(x)$ and $E^u(x)$, $x \in Y$, which satisfy the following conditions:

(1) The functions $\lambda, \mu,$ and ε are f-invariant and

$$\lambda(x) < \lambda(x)e^{\varepsilon(x)} < \mu(x)e^{-\varepsilon(x)} < \mu(x)$$

$$\lambda(x) < 1 < \mu(x);$$

(2) $E^s(x)$ and $E^u(x)$ depend measurably on x and

$$T_x M = E^s(x) \oplus E^u(x),$$

$$Df(x) \cdot E^\tau(x) = E^\tau(f(x)) \quad \text{for} \quad \tau = s, u;$$

(3) if $v \in E^u(f(x))$ and $n \le 0$,

$$\|Df^n \cdot v\| \le C(x)\mu(x)^n e^{\varepsilon(x)|n|} \|v\|;$$

(4) if $v \in E^s(f(x))$ and $n \ge 0$,

$$\|Df^n \cdot v\| \le C(x)\lambda(x)^n e^{\varepsilon(x)n} \|v\|;$$

(5) $\angle(E^u(x), E^s(x)) \ge K(x)$;
(6) for $n \in \mathbb{Z}$,

$$C(f^n(x)) \le C(x)e^{\varepsilon|n|} \quad \text{and} \quad K(f^n(x)) \ge K(x)e^{-\varepsilon|n|}.$$

We say that f is non-uniformly hyperbolic with respect to an invariant measure ν if there exists a Borel set Y with $\nu(Y) = 1$ and such that f is non-uniformly hyperbolic on Y.

Definition 5.2 Let $f : M \to M$ be a $C^{1+\alpha}$ diffeomorphism preserving a probability measure ν on M. We say that ν is a hyperbolic measure , or that f

has non-zero Lyapunov exponents (with respect to ν), if for ν-almost every point x we have

$$\lambda_i(x) \neq 0, \quad 1 \leq i \leq k(x)$$

where $\lambda_i(x), 1 \leq i \leq k(x)$, are the Lyapunov exponents of f at x (see Theorem 5.1).

Observe that a diffeomorphism $f : M \to M$ is non-uniformly hyperbolic with respect to a probability invariant measure ν if and only if ν is a hyperbolic measure. When this is the context and ν is a smooth measure Y. Pesin showed that the Kolmogorov property is equivalent to the Lebesgue measure.

Theorem 5.3 ([25]) *Let f be a $C^{1+\alpha}$ diffeomorphism of a smooth compact Riemannian manifold M preserving an absolutely continuous hyperbolic measure ν. If f is a Kolmogorov automorphism, then it is Bernoulli.*

Comments on the Proof Observe that Pesin's theory provides, for the non-uniformly hyperbolic case, essentially the same kind of geometric invariant structures that are present in the Anosov case. Also, as remarked above, Pesin's also showed that the stable and unstable foliations of non-uniformly hyperbolic diffeomorphisms are absolutely continuous. Therefore, it is not unexpected that the proof of Theorem 5.3 follows the same approach done for the Anosov case in Chap. 4. However, there is a very subtle issue that needs to be treated carefully whenever we work with non-uniformly hyperbolic diffeomorphisms: the sizes of the local stable and unstable manifolds. In the Anosov case Theorem 4.1 gives us a uniform inferior bound for the sizes of local stable (resp. unstable) manifolds. The same is not true for non-uniformly hyperbolic diffeomorphisms. For this reason the generalization is not so straightforward.

The approach to bypass this difficult in to work first inside sets where the functions appearing in item (a) of Definition 5.1 are uniformly bounded. These sets are called *regular sets* or *Pesin set* (cf. *Hyperbolic blocks* in [6, p. 302]).

Definition 5.3 Let λ, μ, and ε be positive numbers satisfying

$$0 < \lambda e^{\varepsilon} < \mu e^{-\varepsilon}, \quad \lambda < 1 < \mu \tag{5.1}$$

and f a non-uniformly hyperbolic diffeomorphism on a full ν-measure set $Y \subset M$ where ν is an f-invariant probability measure. Consider the functions $\lambda(x), \mu(x), C(x)$ and $\varepsilon(x)$ from the definition of non-uniform hyperbolicity. Given an integer j, $1 \leq j \leq n$ and $l \geq 1$, the regular set $\Lambda^l_{\lambda\mu\varepsilon j}$ is defined as the set of all points $x \in Y$ such that:

(1) if $v \in E^u(f(x))$ and $n \leq 0$,

$$||Df^n \cdot v|| \leq l \cdot \mu^n e^{\varepsilon|n|}||v||;$$

(2) if $v \in E^s(f(x))$ and $n \geq 0$,

$$\|Df^n \cdot v\| \leq l \cdot \lambda^n e^{\varepsilon n} \|v\|;$$

(3) $\angle(E^u(x), E^s(x)) \geq l^{-1}$;
(4) $\dim(E^s(x)) = j$.

The union $\Lambda_{\lambda\mu\varepsilon j} = \bigcup_{l\geq 1} \Lambda^l_{\lambda\mu\varepsilon j}$ is called a level set .

It is easy to see that the level set $\Lambda^l_{\lambda\mu\varepsilon j}$ contains the set $Y^l_{\lambda\mu\varepsilon j}$ of all points $x \in M$ such that

$$\lambda(x) \leq \lambda < \mu \leq \mu(x), \varepsilon(x) \leq \varepsilon$$

$$C(x) \leq l, K(x) \geq l^{-1}$$

and $\dim(E^s(x)) = j$. Also we have $Y = \bigcup \Lambda_{\lambda\mu\varepsilon j}$ where the union is taken over all numbers λ and μ satisfying (5.1) and over all $\varepsilon > 0$ and $j \geq 1$.

Observe that level sets are invariant though regular sets need not to be. It was showed by Pesin (see Propositions 2.2.9 from [4]) that regular sets $\Lambda^l = \Lambda^l_{\lambda\mu\varepsilon j}$ are closed and the subspaces $E^s(x)$ and $E^u(x)$ (and consequently the discs $W^s_{loc}(x)$ and $W^u_{loc}(x)$) vary continuously with $x \in \Lambda^l$. Consequently the sizes of $W^s_{loc}(x)$ and $W^u_{loc}(x)$ are uniformly bounded away from zero on each $x \in \Lambda^l$ and so is the angle between the two subspaces.

Returning to the statement of Theorem 5.3, since ν is an ergodic measure, the dimension of $E^s(x)$ as well as the f-invariant functions $\lambda, \mu, \varepsilon$ are constant almost everywhere. Thus there exists a level set Λ with full ν-measure and we can denote its associated regular sets by $\Lambda^l, l \geq 1$, i.e., omitting the constants $\lambda, \mu, \varepsilon$.

With this machinery in hands the proof of Theorem 5.3 is now done by first taking partitions of Λ^l by rectangles and proving, restricted to Λ^l, results that parallel Lemmas 4.8–4.10. As $\nu(\Lambda^l) \to 1$ the result follows. We encourage the reader to read the proof in full details in [4].

5.2 Anosov and Nonuniformly Hyperbolic Flows

The concepts defined in Sect. 5.1 for discrete dynamical systems also exist for continuous dynamics, i.e., dynamical systems where the orbit of a point is not given by iterates of a single transformation but, instead, it is given by letting a parameter vary in a continuum space such as $\mathbb{R}$. Below we make this concept more precise by introducing the notion of a flow on a Riemannian manifold. Along this section several notions from Riemannian-geometry are necessary such as the definition of geodesics and sectional curvature. These concepts can be learned from a textbook on Riemannian-geometry such as [11].

Let M be a Riemannian manifold endowed with a probability measure ν.

Definition 5.4 A C^r flow on $M, r \geq 0$, is a C^r function $\varphi : \mathbb{R} \times M \to M$ satisfying

$$\varphi(0, x) = x, \quad \text{and} \quad \varphi(s, \varphi(t, x)) = \varphi(t + s, x)$$

for all $x \in M, t, s \in \mathbb{R}$. We usually denote $\varphi(t, \cdot) : M \to M$ by φ_t.

We say that a flow φ on M preserves the measure ν if ν is φ_t-invariant for every $t \in \mathbb{R}$.

The definition of uniform hyperbolicity for the discrete dynamics and the existence of stable and unstable foliations for such can be reformulated for flows without much change.

Definition 5.5 A C^1 flow $\varphi : \mathbb{R} \times M \to M$ on a smooth manifold M is said to be an Anosov flow if there exists a Riemannian metric on M and constants $\lambda < 1 < \mu$ such that for all $x \in M$ there is a decomposition

$$T_x M = E^s(x) \oplus E^0(x) \oplus E^u(x)$$

such that

(1) $E^0(x) = \mathbb{R} \cdot \frac{d}{dt}\varphi_t(x)\big|_{t=0}$, with $\frac{d}{dt}\varphi_t(x)\big|_{t=0} \neq 0$;

(2) $D\varphi_t(x)E^\tau(x) = E^\tau(\varphi_t(x))$ for all $t \in \mathbb{R}, \tau = s, u$;

(3) for $t \geq 0$

$$\|D\varphi_t\big|E^s(x)\| \leq \lambda^t, \quad \text{and} \quad \|D\varphi_{-t}\big|E^u(x)\| \leq \mu^{-t}.$$

One of the most important examples of a flow on a Riemannian manifold is the so-called *geodesic flow*. Let M be a complete Riemannian manifold and $T^1 M$ its unit tangent bundle

$$T^1 M = \{(x, v) : x \in M, v \in T_x M \quad \text{and} \quad \|v\| = 1\}.$$

For any $v \in T_x M$ with $\|v\| = 1$ there is a unique geodesic $\gamma_v(t)$ such that

$$\gamma_v(0) = x \quad \text{and} \quad \frac{d}{dt}\gamma_v(t)\big|_{t=0} = v. \tag{5.2}$$

The geodesic flow is the flow $g : \mathbb{R} \times T^1 M \to T^1 M$ given by

$$g_t(x, v) = \left(\gamma_v(t), \frac{d}{dt}\gamma_v(t)\right),$$

where $\gamma_v(t)$ is the unique geodesic satisfying (5.2).

Theorem 5.4 *Let M be a compact Riemannian manifold with negative sectional curvature. The geodesic flow g_t on M is an Anosov flow.*

Proof See [17, Sections 17.5 and 17.6]. □

The stable manifold theorem can also be extended for the case of Anosov flows as the following result shows.

Theorem 5.5 *[17, Theorem 17.4.3] Let $\varphi : \mathbb{R} \times M \to M$ be a C^r Anosov flow on a compact smooth manifold M and let λ, μ be as in Definition 5.5. Let $t_0 > 0$ be a fixed real number. Then for each $x \in M$ there is a pair of embedded disks C^r-disks $W^s(x)$, $W^u(x)$ called the local strong stable manifold and the strong unstable manifold at x, respectively, such that*

(1) $T_x W^s(x) = E^s(x), T_x W^u(x) = E^u(x)$;
(2) $\varphi_t(W^s(x)) \subset W^s(\varphi_t(x))$ and $\varphi_{-t}(W^u(x)) \subset W^u(\varphi_{-t}(x))$ for $t \geq t_0$;
(3)   for every $\delta > 0$ there exists $C(\delta)$ such that

$$d(\varphi_t(x), \varphi_t(y)) < C(\delta)(\lambda + \delta)^t d(x, y) \quad \text{for} \quad y \in W^s(x), t > 0,$$

$$d(\varphi_{-t}(x), \varphi_{-t}(y)) < C(\delta)(\mu - \delta)^{-t} d(x, y) \quad \text{for} \quad y \in W^u(x), t > 0,$$

(4)   there exists a continuous family U_x of neighborhoods of $x \in M$ such that

$$W^s(x) = \{y \mid \varphi_t(y) \in U_{\varphi_t(x)}, \ t > 0, \ and \ d(\varphi_t(x), \varphi_t(y)) \to 0 \ as \ t \to \infty\},$$

$$W^u(x) = \{y \mid \varphi_{-t}(y) \in U_{\varphi_{-t}(x)}, \ t > 0, \ and \ d(\varphi_{-t}(x), \varphi_{-t}(y)) \to 0 \ as \ t \to \infty\}.$$

For a brief comment on how to prove Theorem 5.5 we refer the reader to Section 17.4 of [17].

Definition 5.6 Let (X, ν) be a Lebesgue space. A measure preserving flow $\varphi : \mathbb{R} \times X \to X$ is said to be a K-flow (resp. Bernoulli flow) if the time-one map $\varphi_1 : X \to X$ is a Kolmogorov automorphism (resp. a Bernoulli automorphism). When a flow is a K-flow (resp. a Bernoulli flow) we also say that φ has the Kolmogorov property (resp. the Bernoulli property).

Theorem 5.6 ([7]) *The geodesic flow on a compact C^2-manifold of negative sectional curvature is a Bernoulli flow.*

The proof of Theorem 5.6 was given by Ornstein and Weiss using the same lines as their proof of the same result for linear automorphisms of $\mathbb{T}^2$. In the proof it is necessary to use the Kolmogorov property of such flows which was established earlier in the works of Anosov [1, 2].

Similar to what happens for the discrete case, the concept of uniform hyperbolicity for flows can be weakened to the concept of nonuniform hyperbolicity for flows by allowing some flexibility in the constants appearing in Definition 5.5 (see [4] for the definition). Also, the concept of Lyapunov exponents and, consequently, the concept of hyperbolic invariant measure of a flow can be generalized as follows:

Theorem 5.7 (Oseledets Theorem for Continuous Dynamics) *Let $\varphi : \mathbb{R} \times M \to M$ be a C^1 flow defined on a compact Riemannian manifold M. The set of points $x \in M$ which satisfies:*

- *there exists a splitting*

$$T_x M = E_1(x) \oplus \ldots \oplus E_{k(x)};$$

- *$D\varphi_t(x)E_i(x) = E_i(\varphi_t(x))$, $k(x) = k(\varphi_t(x))$, for all $t \in \mathbb{R}$;*
- *there exist the limits*

$$\lambda_i(x) := \lim_{t \to \infty} \frac{1}{t} \log \|D\varphi_t(x) \cdot v\| = \lim_{t \to -\infty} \frac{1}{t} \log \|D\varphi_t(x) \cdot v\|,$$

for all $v \in E_i(x) \setminus \{0\}$,

is a set of full measure for any invariant probability measure for φ. The numbers $\lambda_i(x)$ are called the Lyapunov exponents of the flow φ at x.

Definition 5.7 Let $\varphi : \mathbb{R} \times M \to M$ be a C^1 flow preserving a probability measure ν on M. We say that ν is a hyperbolic measure if for ν-almost every point x we have

$$\lambda_i(x) \neq 0, \quad 1 \leq i \leq k(x)$$

where $\lambda_i(x)$, $1 \leq i \leq k(x)$, are the Lyapunov exponents of φ at x (see Theorem 5.7).

For flows preserving a hyperbolic measure one can prove a version of Theorem 5.5 but with the same technical issue on the non-uniform bound on the sizes of local stable and unstable manifolds (as remarked in the comments of the proof of Theorem 5.3 for the discrete case). To treat this difficulty Pesin also worked with a version of *level* and *regular* sets for nonuniformly hyperbolic flows where the sizes of local stable and local unstable manifolds can be taken to be uniformly bounded away from zero. Using techniques which are very similar to the proof of Theorem 5.3, Pesin proved that the Kolmogorov and the Bernoulli property are also equivalent for nonuniformly hyperbolic flows.

Theorem 5.8 ([24, 26]) *Assume that the flow φ_t preserves an absolutely continuous hyperbolic measure ν and has the Kolmogorov property. Then φ_t is a Bernoulli flow.*

5.3   Nonuniformly Hyperbolic Maps and Flows with Singularities

In 1986 Pesin's theory was extended to the much more general setting of smooth maps with singularities by Katok and Strelcyn in the fundamental monograph [18]. One of the motivations to develop this theory is that this class of functions, which we will quickly define below, includes important dynamical systems occurring in classical mechanics as, for example, the motion of systems of rigid balls with elastic collisions and, consequently, is an efficient tool to study the behavior of the so-called billiard systems. The problem of obtaining the Bernoulli property for Billiards was studied by Gallavotti and Ornstein in [13] where the authors showed that the billiard flow associated to a dispersing billiard table with smooth boundary on the two-torus and its associated map on the boundary are Bernoulli. In 1981 Kubo and Murata [19] extended this result to small perturbations of such billiard flows. In [18] the authors remark that the estimate they obtain is not enough to obtain Bernoulli property from the Kolmogorov property for the class they are working with. This problem was solved only two decades later by Chernov and Haskell in [10] where, in particular, the authors showed that several types of Billiard systems (which were previously known to be Kolmogorov) are also Bernoulli. In what follows we briefly present Chernov–Haskell results without entering in much details about billiards systems and we refer the reader to [10] for a detailed discussion on the subject and definitions.

First let us describe the general setting in which the results will be stated. Let M be a smooth compact d-dimensional Riemannian manifold (possibly with boundary ∂M). Let Γ be a closed subset of M and denote $S_1 := \partial M \cup \Gamma$. Let $T : M \setminus S_1 \to M$ be a C^2 diffeomorphism of $M \setminus S_1$ onto its image and denote

$$S_{-1} := M \setminus T(M \setminus S_1).$$

We refer to S_1 as the singularity set for T and S_{-1} the set of singularities for T^{-1}.

Definition 5.8 Let ρ be the Riemannian metric on M, denote by $B_r(x)$ the ball of radius r centered at $x \in M$ and by $O_r(A)$ the r-neighborhood of $A \subset M$, that is,

$$O_r(A) = \bigcup_{x \in A} B_r(x).$$

Let ν be a T-invariant absolutely continuous probability measure on M. Assume that the following are true:

(1) for some constants $a_1, c_1 > 0$ we have

$$\nu(O_\varepsilon(S_1 \cup S_{-1})) \leq c_1 \varepsilon^{a_1},$$

  in particular $\nu(S_1 \cup S_{-1}) = 0$;

(2) for some constants $a_2, c_2 > 0$

$$\|D^2 T(x)\| \leq c_2 \rho(x, S_1)^{-a_2} \quad \text{and} \quad \|D^2 T^{-1}(x)\| \leq c_2 \rho(x, S_{-1})^{-a_2}$$

  for every $x \in M$;

(3)

$$\int_M \ln^+ ||DT(x)||dv < \infty \quad \text{and} \quad \int_M \ln^+ ||DT^{-1}(x)||dv < \infty,$$

where $\ln^+(x) = \max\{\ln(x), 0\}$.

Then T is called a smooth map with singularities.

The setting of Definition 5.8 guarantees the applicability of the Oseledec Theorem (see Theorem 5.1) and, consequently, ensures the existence of Lyapunov exponents almost everywhere in M. If all the Lyapunov exponents are non-zero, then T is called *completely hyperbolic*.

Theorem 5.9 ([10]) *Let (M, T, v) be a smooth system with singularities satisfying the conditions of Definition 5.8. If the map T is completely hyperbolic and Kolmogorov, then it is Bernoulli.*

As usual, one can make a parallel definition for the continuous case. Here such definition is made by taking a suspension flow over a smooth map with singularities T and C^2 positive integrable function φ defined on $M \setminus S_1$.

Definition 5.9 ([10]) Let (M, T, v) be a nonuniformly hyperbolic map with singularities as in Definition 5.8 and v be a smooth invariant measure. Let $\varphi(x)$ be a positive integrable C^2-function on $M \setminus S_1$. A suspension flow with a base map T and a ceiling function φ is defined on the manifold $\mathcal{M} := \{(x, s) : x \in M, 0 \leq s \leq \varphi(x)\}$ by

$$\Phi^t(x, s) = (x, s + t) \quad \text{for } 0 \leq t \leq \varphi(x) - s$$

$$\Phi^t(x, s) = (Tx, s + t - \varphi(x)) \quad \text{for } \varphi(x) - s \leq t \leq \varphi(Tx) + \varphi(x) - s.$$

This flow preserves the smooth probability measure μ on M given by $d\mu = c \cdot dv \times ds$ where $c^{-1} := \int_M \varphi(x)dv(x)$. It is clear that the Lyapunov exponent of the tangent vector to the flow is zero. If all the other Lyapunov exponents are non-zero, then Φ^t is said to be completely hyperbolic.

Theorem 5.10 ([10]) *Let $(\mathcal{M}, \Phi^t, \mu)$ be a suspension flow over a smooth map T with singularities satisfying Definition 5.8 and with C^2 ceiling function. If the flow Φ^t is completely hyperbolic and Kolmogorov, then it is Bernoulli.*

As a consequence of these theorems several types of systems which were previously shown to be Kolmogorov are actually Bernoulli. Among those are the semidispersing planar billiards, the planar billiards with focusing components of special types, the systems of hard balls and the model of the periodic Lorentz gas (see [10] for definitions). We remark that the conditions stated in Definition 5.8 guarantee the existence of local unstable and local stable manifolds which are at least C^1 and which are also absolutely continuous (see [18]). Using these invariant structures in [10] the ideas for the proofs of both theorems above are still based on the same philosophy established by the Ornstein–Weiss argument but adding some technicalities to deal with the sets of singularities.

5.4 Partially Hyperbolic Diffeomorphisms

So far, all the cases we have seen include some kind of hyperbolicity, even in the case of completely hyperbolic systems with singularities. We now introduce a class of maps which may present, in some direction, no hyperbolicity at all.

Definition 5.10 Given a smooth compact Riemannian manifold M. A diffeomorphism $f : M \to M$ is called partially hyperbolic if the tangent bundle of the ambient manifold admits an invariant decomposition $TM = E^s \oplus E^c \oplus E^u$, such that all unit vectors $v^\sigma \in E_x^\sigma, \sigma \in \{s, c, u\}$ for any $x, y, z \in M$

$$\|D_x f v^s\| < \|D_y f v^c\| < \|D_z f v^u\|$$

and $\|D_x f v^s\| < 1 < \|D_z f v^u\|$ where v^s, v^c, and v^u belong, respectively, to E_x^s, E_y^c, and E_z^u.

This is also referred to as absolute partially hyperbolic to distinguish from the definition of pointwise partial hyperbolicity which has the following condition instead of the first relation above: for any $x \in M$

$$\|D_x f v^s\| < \|D_x f v^c\| < \|D_x f v^u\|.$$

As for the non-uniformly hyperbolic case, it is well known that for partially hyperbolic diffeomorphisms there are foliations $\mathscr{F}^\tau, \tau = s, u$, tangent to the subbundles $E^\tau, \tau = s, u$, called *stable* and *unstable foliation* respectively (for more details see, for example, [4]). On the other hand, the integrability of the central subbundle E^c is a subtle issue and is not the case for a general partially hyperbolic diffeomorphism (see [15]). However, a result of Brin et al. [8] guarantees that on $\mathbb{T}^3$ all absolute partially hyperbolic diffeomorphisms admit a central foliation tangent to the direction E^c.

Also, as for the case of Anosov diffeomorphisms (see Theorem 4.4), the stable and unstable foliations of a partially hyperbolic diffeomorphism are absolutely continuous (see Definition 4.6) and, furthermore, the holonomy map between two transversals is absolutely continuous with Jacobian going to one as the transverses approach each other. (see [4, 7, 24, 29, 30]).

Let $f : \mathbb{T}^n \to \mathbb{T}^n$ be a partially hyperbolic diffeomorphism. Consider $f_* : \mathbb{Z}^n \to \mathbb{Z}^n$ the action of f on the fundamental group of $\mathbb{T}^n$. f_* can be extended to $\mathbb{R}^n$ and the extension is the lift of a unique linear automorphism $A : \mathbb{T}^n \to \mathbb{T}^n$.

Definition 5.11 Given $f : \mathbb{T}^n \rightarrow \mathbb{T}^n$ a partially hyperbolic diffeomorphism. The unique linear automorphism $A : \mathbb{T}^n \rightarrow \mathbb{T}^n$ with lift $f_* : \mathbb{R}^n \rightarrow \mathbb{R}^n$, as constructed in the previous paragraph, is called the linearization of f.

Definition 5.12 We say that $f : \mathbb{T}^n \rightarrow \mathbb{T}^n$ is derived from Anosov diffeomorphism, or just a DA diffeomorphism, if it is partially hyperbolic and its linearization is a hyperbolic automorphism (no eigenvalue has norm one).

In [28] the authors obtained a condition in terms of the measurable properties of the center-stable/center-unstable foliations which promote the Kolmogorov property to the Bernoulli property.

Theorem 5.11 ([28]) *Let* $f : \mathbb{T}^3 \rightarrow \mathbb{T}^3$ *be a* C^2 *volume preserving derived from Anosov diffeomorphism with linearization* $A : \mathbb{T}^3 \rightarrow \mathbb{T}^3$. *Assume that* f *is Kolmogorov and one of the following occurs:*

1. $\lambda_A^c < 0$ *and* $\mathscr{F}^{cs}$ *is absolutely continuous, or*
2. $\lambda_A^c > 0$ *and* $\mathscr{F}^{cu}$ *is absolutely continuous.*

Then f *is Bernoulli.*

It is still not clear if absolute continuity of the center manifold implies the absolute continuity of $\mathscr{F}^{cs}$ and of $\mathscr{F}^{cu}$. We also emphasize that the absolute continuity condition is essential to prove Bernoulli property of smooth measures in our setting and in all other contexts cited before.

In a recent paper Kanigowski et al. [16] constructed examples on $\mathbb{T}^4$ of volume-preserving Kolmogorov partially hyperbolic diffeomorphisms which are not Bernoulli. We highly recommend that the reader read their article where they not only construct the mentioned example but also discuss on previous examples of Kolmogorov but not Bernoulli maps (for both the smooth and non-smooth context). In particular, they exhibit a very instructive chart comparing properties satisfied for several of the examples appearing in the literature.

Although it seems a difficult problem, we conjecture that there may exist Kolmogorov but not Bernoulli partially hyperbolic diffeomorphisms in three-dimensional manifold.

Conjecture 5.1 ([28]) There exist Kolmogorov but not Bernoulli partially hyperbolic diffeomorphisms in three-dimensional manifolds.

The main novelty of the proof of Theorem 5.11 is that it introduces a new approach to the problem by using disintegration of measures along the center foliation to overcome the absence of hyperbolicity along the center direction. In what follows we describe deeper some of the crucial aspects of this approach.

5.4.1 Virtual Hyperbolicity for DA Diffeomorphisms of $\mathbb{T}^3$

If $f : \mathbb{T}^3 \to \mathbb{T}^3$ is a DA diffeomorphism, then by results of Franks [12] and Manning [22] there is a semi-conjugacy $h : \mathbb{T}^3 \to \mathbb{T}^3$, which we will call the Franks-Manning semi-conjugacy, between f and its linearization A, that is, h is a continuous surjection satisfying

$$A \circ h = h \circ f \tag{5.3}$$

Moreover, this semi-conjugacy has the property that there exists a constant $K \in \mathbb{R}$ such that if $\tilde{h} : \mathbb{R}^3 \to \mathbb{R}^3$ denotes the lift of h to $\mathbb{R}^3$ we have $\|\tilde{h}(x) - x\| \leq K$ for all $x \in \mathbb{R}^3$ and, given two points $a, b \in \mathbb{R}^3$, there exists a constant $\Omega > 0$ with

$$\tilde{h}(a) = \tilde{h}(b) \Leftrightarrow \|\tilde{f}^n(a) - \tilde{f}^n(b)\| < \Omega, \forall n \in \mathbb{Z}, \tag{5.4}$$

where $\tilde{f}$ denotes the lift of f to $\mathbb{R}^3$. In [33] Ures proved that h maps center leaves of f onto center leaves of A, i.e., $\mathscr{F}_A^c(h(x)) = h(\mathscr{F}_f^c(x))$, and, as a consequence, he proved that given any point $x_0 \in \mathbb{T}^3$ the set $h^{-1}(\{x_0\})$ is uniformly bounded connected set (i.e., a segment) inside the center foliation. In other words, if h is not injective, then the sets where injectivity fails are arcs inside center leaves. Given a point $x \in \mathbb{T}^3$ define the set $c(x) \subset \mathscr{F}^c(x)$ by: $c(x) := h^{-1}(\{h(x)\})$. By the above discussion, the diameter of $c(x)$ is uniformly bounded in x. Take

$$\mathscr{C} := \bigcup_{y \in \{x \in \mathbb{T}^3 \mid c(x) \neq \{x\}\}} c(y), \tag{5.5}$$

in other words $\mathscr{C}$ is the set of all non-degenerated arcs inside center leaves which are collapsed to a point by h. It is easy to see that $f(\mathscr{C}) = \mathscr{C}$.

Definition 5.13 An f-invariant measure μ is called virtually hyperbolic if there exists a full measurable invariant subset Z such that Z intersects each center leaf $\mathscr{F}^c(x)$, $x \in M$, in at most one point.

The above definition was given in [21] in the context of algebraic automorphisms and the existence of such measures in partially hyperbolic diffeomorphism also had been noticed before (see, for instance,[27, 32]). If μ is virtually hyperbolic , then the central foliation is measurable with respect to μ and conditional measures along center leaves are (mono-atomic) Dirac measures. Indeed the partition into central leaves is equivalent to the partition into points.

As we have seem in the previous sessions, to obtain the Bernoulli property for systems with a hyperbolic structure, some features of hyperbolic systems are fundamental, these are: the existence of a pair of foliations $\mathscr{F}^s$, $\mathscr{F}^u$ which are, respectively, contracting and expanding, transversal to each other, and absolutely continuous. With this structure in hand, the standard procedure is to take a partition (usually a partition where each element has piecewise-smooth boundaries) and

prove that it is Very Weak Bernoulli. As a consequence of Ornstein Theory it follows
that the system is Bernoulli.

In the literature, the Bernoulli property is also obtained via Ornstein Theory
arguments when we have some type of symbolic dynamics associated to the
dynamical system under consideration. This is the case when we have Markov
partitions or some type of Markov structure for the dynamical system. It was
using this type of approach that Ratner [31] proved that Anosov flows with u-
Gibbs measures are Bernoulli. Also using a symbolic approach, Ledrappier et al.
[20] proved that, with respect to an ergodic measure of maximum entropy, smooth
flows with positive speed and positive topological entropy on a compact smooth
three-dimensional manifold are either Bernoulli or isomorphic to the product of a
Bernoulli flow and a rotational flow.

The fundamental difficulty in the partially hyperbolic context is that we lack
hyperbolic behavior on the center direction and we do not have, a priori, any kind of
Markov partitions (thus we lack a symbolic representation for the system). When we
restrict to the context derived from Anosov diffeomorphisms (see Definition 5.12)
we have the advantage that the center foliation in some sense carries some information from the center foliation of its linearization. If the linearization of a derived
from Anosov diffeomorphism f has negative center exponent for example, then the
center foliation of f has globally the same behavior (expansion or contraction) as
the center foliation for the linearization.

In view of this fact, the idea to tackle the problem derived from Anosov diffeomorphisms is to treat the center foliation as a contracting (or expanding) foliation
in "as many points as possible." This idea is the key to proof of Theorem 3.6. Given
a derived from Anosov diffeomorphism f with linearization A, we have the semi-
conjugacy h between f and A defined above. Assume that the center Lyapunov
exponent of A is negative. We prove (see Lemma 5.7) that if a pair of points
$(a, b) \in M \times M$ is such that $h(a) \neq h(b)$ and $b \in \mathscr{F}^{cs}(a)$ then their orbits by
f behaves, for most of the time, as if they were in a same contracting foliation,
that is, the distance between $f^n(a)$ and $f^n(b)$ is very small for most of the natural
numbers n. Thus, if we restrict our analysis to the set of points $x \in M$ for which
$h(y) \neq h(x)$ for all $y \in M \setminus \{x\}$, we can "treat" the center foliation essentially as
if it was a contracting foliation. Although, we have no reason to assume at first that
this set has total measure. Therefore the first question that should be addressed is:
how large is the set of points which are in the injectivity domain of h, that is, how
large is the set

$$\{x \in M : h(y) \neq h(x) \text{ for all } y \in M \setminus \{x\}\}?$$

In Theorem 5.12 we show that if this set has zero measure then there exists a
full measure set intersecting almost every center leaf in exactly one point. We call
such set as the set of atoms. Since points in two separate center leaves have distinct
images by h then we can restrict our analysis to the set of atoms and again treat, in
some sense, the center foliation as a "contracting foliation."

Several technical issues appear when we execute the idea outlined above. One of the technical issues is that when dealing with the injectivity domain of h, we need to prove that we are making measurable choices of sets. For example, we need to prove that the partition by collapsed pieces (by the sets $c(x)$) is indeed a measurable partition. Later we need to apply that h maps the union of collapsed pieces to a measurable set (see Lemma 5.2) and use it, together with a Measurable Choice Theorem (Theorem 5.13), to prove that the extreme bottom points of collapsed arcs $c(x)$ around x varies measurably on $x \in M$.

Theorem 5.12 (Ponce–Tahzibi–Varão [28]) *Let $f : \mathbb{T}^3 \to \mathbb{T}^3$ be a C^2 volume (m) preserving derived from Anosov diffeomorphism with h a semi conjugacy to linear Anosov diffeomorphism. Then, h is $m-$almost everywhere injective. More precisely, the following dichotomy is valid:*

- *Either the set $\mathscr{C}$ has zero volume, or*
- *$\mathscr{C}$ has full measure and (f, m) is virtually hyperbolic.*

In the latter case, (f, m) is Kolmogorov.

It is clear that h is injective in $\mathbb{T}^3 \setminus \mathscr{C}$ and if $m(\mathscr{C}) = 1$ then, by second part of the dichotomy of Theorem 5.12, there exists a full measurable subset (of atoms) which intersects each leaf in exactly one point and h restricted to this subset is injective. This motivates the following definition of "essential injectivity domain" of h.

Definition 5.14 The essential injectivity domain X of h is defined as follows: If $m(\mathscr{C}) = 0$, we take $X := \mathbb{T}^3 \setminus \mathscr{C}$. Otherwise we define $X =$ set of atoms.

Theorem 5.12 is interesting on its own, as it shows the existence of an isomorphism between the volume measure and its image under the semi conjugacy. We recall a result of Buzzi and Fisher [9] where they prove entropic stability for a class of deformation of Anosov automorphisms. Although it is not written explicitly, their method can be applied for all derived from Anosov diffeomorphisms to obtain isomorphism between f and its linearization for high entropy measures. In our case, since we are working specifically with the volume measure which has not necessarily high entropy, their arguments cannot be applied.

Remark 5.1 Theorems 5.11 and 5.12 admit a generalization for higher dimensional cases under some conditions on the center manifold and on the lift of the stable and unstable foliations. We refer the reader to [28] for a discussion on such generalizations.

5.4.2 Proof of Theorem 5.12: Virtual Hyperbolicity of Lebesgue Measure

In order to prove Theorem 5.12 we need to use, what is known as the Measurable Choice Theorem. This result, proved by Aumann [3], has an intrinsic interest from

the mathematical point of view, although it has appeared in the context of Decision
Theory from Economics.

Theorem 5.13 (Measurable Choice Theorem [3]) *Let (T, μ) be a σ-finite
measure space, let S be a Lebesgue space, and let G be a measurable subset of $T \times S$
whose projection on T is all of T. Then there is a measurable function $g : T \to S$,
such that $(t, g(t)) \in G$ for almost all $t \in T$.*

Proof of Theorem 5.12 First of all let us mention that the dichotomy in the Theorem
B immediately implies that h is almost everywhere injective. Indeed, if $m(\mathscr{C}) = 0$,
then h is injective outside $\mathscr{C}$ which has full measure, and if $m(\mathscr{C}) = 1$ then we
prove that m is virtually hyperbolic or in other words there exists a full measurable
subset intersecting central leaves in at most one point and consequently h restricted
to this subset is injective.

So let us prove the dichotomy. If the Lebesgue measure of $\mathscr{C}$ does not vanish,
then we claim that f is accessible, otherwise by Hammerlindl and Ures [14] it
would be topologically conjugate to its linearization, hence the set $\mathscr{C} = \emptyset$, which
is an absurd. Thus f is accessible, therefore Kolmogorov. In particular f and f^2
are ergodic, then we can assume without loss of generality that f preserves the
orientation of $\mathscr{F}^c$ (otherwise we work with f^2). Since $\mathscr{C}$ is an invariant set with
positive volume, it has full volume.

Now we need to prove that m is virtually hyperbolic. To do this, we restrict m to
the full measure set $\mathscr{C}$ and consider its disintegration along the partition by intervals
$c(x), x \in \mathscr{C}$:

Lemma 5.1 *If $m(\mathscr{C}) = 1$, then the partition $\mathscr{C} = \{c(x) : x \in \mathscr{C}\}$ of $\mathbb{T}^3$ by
collapsed arcs of center manifolds is a measurable partition.*

Proof of Theorem 5.12 Consider $\{B_i\}_{i \in \mathbb{N}}$ a countable basis of open sets on $\mathbb{T}^3$.
Then, we know that each point $x_0 \in \mathbb{T}^3$ can be written as an intersection
$\{x_0\} = \bigcap B_i^*$, where $B_i^* = B_i$ or B_i^c. Thus,

$$h^{-1}(\{x_0\}) = \bigcap h^{-1}(B_i^*).$$

Since h is continuous, $h^{-1}(B_i^*)$ is Lebesgue measurable (since it is an open or a
closed set). The countable family $\{h^{-1}(B_i^*)\}$ separates the sets $h^{-1}(\{x_0\})$. Thus $\mathscr{C}$
is a measurable partition.   □

Let $\{m_{c(x)}\}_{c(x) \in \mathscr{C}}$ be the Rokhlin disintegration of volume m on the partition
$\mathscr{C}$. Define $\eta := h_* m$, notice that η is an invariant measure for A (recall that A is
the linearization of f). And because the collapsed intervals have full measure by
hypothesis, then η is an atomic and invariant measure for the linear map A. We now
use the following result from [27].

Theorem 5.14 ([27]) *Let $f : \mathbb{T}^3 \to \mathbb{T}^3$ be a volume preserving, DA diffeomorphism. Suppose its linearization A has the splitting $T_x M = E^{su}(x) \oplus E^{wu}(x) \oplus
E^s(x)$ (su and wu represents strong unstable and weak unstable.) If f has $\lambda^c(x) < 0$*

for Lebesgue almost every point $x \in \mathbb{T}^3$, then volume has atomic disintegration on $\mathscr{F}_f^c$, in fact the disintegration is mono atomic.

We apply Theorem 5.14 to η and obtain that η has atomic disintegration and one atom per leaf. Although Theorem 5.14 deals with volume measure we point out that the same argument[1] works *ipsis litteris* by changing volume to η.

One atom per center leaf of the disintegration of η implies that for m almost every x the disintegration $\{m_{c(x)}\}_{c(x) \in \mathscr{C}}$ satisfies: if $c(x) \neq c(y)$, then $c(x)$ and $c(y)$ are in distinct center leaves.

Observe that up to now, we get a full Lebesgue measurable subset which intersects almost all leaves at most in a unique interval (a collapse interval $c(x)$). A priori the conditional measure of m supported on $c(x)$ may be non-atomic. In the sequel we prove that this is not the case and indeed such conditional measures are Dirac measures. This implies virtual hyperbolicity of m.

First of all let us prove the measurability of the set $h(\mathscr{C})$.

Lemma 5.2 *The set*

$$h(\mathscr{C}) = \{x \in \mathbb{T}^3 : \#\{h^{-1}(x)\} \neq 1\}$$

is a measurable set, in fact a Souslin set, where $\#\{h^{-1}(x)\}$ means the cardinality of the set $\{h^{-1}(x)\}$.

Proof Let us denote $\tilde{f}$ and $\tilde{h}$, respectively, as the lift of f and h to the universal cover $\mathbb{R}^3$. As pointed out above $\tilde{h}(x) = \tilde{h}(y)$ if and only if $\|\tilde{f}^n(x) - \tilde{f}^n(y)\| \leq \Omega$, $\forall n \in \mathbb{Z}$, where Ω is some constant which only depends on f. Hence, we may characterize the points where $\tilde{h}$ is injective as the following:

$$\#\{\tilde{h}^{-1}(\{x\})\} = 1 \text{ iff } x = \bigcap_{n \in \mathbb{Z}} \tilde{f}^{-n}(B(\tilde{f}^n(x), \Omega)),$$

where $B(\tilde{f}^n(x), \Omega)) \subset \mathbb{R}^3$ is the ball centered at $\tilde{f}^n(x)$ and radius Ω. Define $B_{\tilde{f}}(x) := \bigcap_{n \in \mathbb{Z}} \tilde{f}^{-n}(B(\tilde{f}^n(x), \Omega))$ and $B_{\tilde{f}}^n(x) := \bigcap_{k=-n}^n \tilde{f}^{-k}(B(\tilde{f}^k(x), \Omega))$.

Consider the set $P := \{(x, y) \in \mathbb{R}^3 \times \mathbb{R}^3 \mid x \in B_{\tilde{f}}(y)\}$, also notice that $P = \{(x, y) \mid \tilde{h}(x) = \tilde{h}(y)\}$, and define $P_n := \{(x, y) \in \mathbb{R}^3 \times \mathbb{R}^3 \mid x \in B_{\tilde{f}}^n(y)\}$. It is trivial to see that the diagonal $\Delta \subset P$, since $\tilde{h}(x) = \tilde{h}(x)$. Hence $P \setminus \Delta$ contains the information of the points for which $\tilde{h}$ is not injective, more precisely: $\pi_1(P) = \tilde{\mathscr{C}}$, where $\pi_1 : \mathbb{R}^3 \times \mathbb{R}^3 \to \mathbb{R}^3$ is the projection onto the first coordinate and $\tilde{\mathscr{C}}$ is the lift of $\mathscr{C}$ to the universal cover.

---

[1] The same argument works if the measure η satisfies the following: if we take a Markov partition $\{R_i\}_{i=1}^l$ of $\mathbb{T}^3$, then we should have $\eta(\partial R_i) = 0$, $1 \leq i \leq l$. The measure η defined by $\eta = h_* m$ satisfies such property.

Because $\tilde{f}$ is a homeomorphism, then P_n is a Borel set. And because $P = \bigcap_{n \in \mathbb{N}} P_n$, then P is a Borel set. Therefore, $\tilde{h}(\mathscr{C}) = \tilde{h} \circ \pi_1(P)$ is a Souslin set by Bogachev [5, Corollary 1.10.9]. Now projecting everything to the torus and because the projection is a local homeomorphism we obtain that $h(\mathscr{C})$ is a Souslin set, in particular a measurable set.   □

Consider $\phi(t, .) : \mathbb{T}^3 \to \mathbb{T}^3$ the flow on $\mathbb{T}^3$ having constant speed one in the center direction for the linearization A. More precisely, we know that the leaves of the center foliation of A are straight lines and orientable by assumption. Define $\phi(t, x)$ the unique point in the $\mathscr{F}_A^c(x)$ which has distance t inside this center leaf and in the positive direction from x.

In particular, the above lemma implies that $\phi(-1/n, h(\mathscr{C}))$ is a measurable set. Consider the set $h^{-1}(\phi(-1/n, h(\mathscr{C})))$ which is a measurable set since h is continuous.

Now consider $\Sigma := \mathscr{C}/\sim$ where the relation is given by: $x \sim y$ iff $x \in c(y)$. Let us define the function $\psi_n : \Sigma \to h^{-1}(\phi(-1/n, h(\mathscr{C}))) \subset \mathbb{T}^3$ to be a function given by the Measurable Choice Theorem 5.13 applied to the product $\Sigma \times \mathbb{T}^3$ and the measurable set we consider is $\{([x], y) \in \Sigma \times \mathbb{T}^3 | h(y) = \phi(-1/n, h(x))\}$. We have to use this theorem because we cannot assure that this set intersects the center leaf in one point, it could, of course, intersect on a segment.

But notice that fixed $[x] \in \Sigma$ then $\psi_n([x])$ is an increasing sequence (recall that we have an orientation for the center direction). Therefore we can define the following function

$$\psi : \Sigma \to \mathbb{T}^3$$

$$[x] \mapsto \lim_{n \to \infty} \psi_n([x])$$

and by its construction $\psi([x])$ is the lower extreme of $c(x)$. We now want to prove that the image $Im(\psi)$ is a measurable set. Notice that ψ is measurable function because it is the limit of measurable functions.

By Lusin's theorem there exists a compact set $K_n \subset \Sigma$ such that $\psi|K_n$ is a continuous function and $\hat{m}(\Sigma \setminus K_n) < 1/n$ where $\hat{m} := \pi_* m$ and $\pi : \mathscr{C} \to \Sigma$ is the projection. Because $\psi|K_n$ is a continuous function $\psi(K_n)$ is a compact set. Now $\hat{m}(\Sigma)(\bigcup_n \psi(K_n)) = \hat{m}(\Sigma)$, without loss of generality we may consider that $\Sigma = \bigcup_n \psi(K_n)$, therefore

$$\psi(\Sigma) = \bigcup_n \psi(K_n) \text{ is a measurable set.}$$

We have proven so far that the base of the intervals from $\mathscr{C}$ forms a measurable set and we call these sets as point zero, that is if $x \in \mathscr{C}$ then 0_x means the base point associated to the segment $c(x)$. Let us denote the set of these "zero" points as Σ_0. Observe that Σ_0 may be identified with Σ and equipped with the quotient measure $\hat{m}$. If $y \in c(x)$, then $[0_x, y]$ stands for the segment inside the center leaf which contains 0_x and y.

Let $W_\epsilon := \{\phi(t, 0_y) \mid 0 \le t \le \epsilon, \ y \in \mathscr{C}\}$. Since $[0, \epsilon] \times \Sigma_0$ is a Souslin set and the flow ϕ is a continuous map, the set W_ϵ is a measurable set by Bogachev [5, Proposition 1.10.8]. Define

$$\mu_\epsilon : \Sigma_0 \to [0, 1] \subset \mathbb{R}$$

$$[y] \mapsto m_y(W_\epsilon),$$

this is a measurable function by Rokhlin's Theorem 2.7. Hence,

$$\mu_\epsilon^{-1}([0, \alpha]) = \{0_y \mid m_y(W_\epsilon) \le \alpha\}$$

is a measurable set. We now consider the only three possible cases:

1. for all $\epsilon > 0$ and all $\alpha \in (0, 1]$, $\hat{m}(\mu_\epsilon^{-1}((0, \alpha])) = 0$;
2. there exists $\epsilon > 0$ and there exists $\alpha \in (0, 1)$ such that $\hat{m}(\mu_\epsilon^{-1}((0, \alpha])) > 0$;
3. there exists $\epsilon > 0$ such that $\hat{m}(\mu_\epsilon^{-1}((0, 1])) > 0$ and $\hat{m}(\mu_\epsilon^{-1}((0, \alpha])) = 0$ for all $\alpha < 1$.

Assume we are in the first case. Observe that since the elements $c(x) \in \mathscr{C}$ have uniformly bounded length, for large enough $\epsilon > 0$ we have $\hat{m}(\mu_\epsilon^{-1}([0, 1])) = 1$. Thus, for a large enough $\epsilon > 0$ we must have $\hat{m}(\mu_\epsilon^{-1}(\{0\})) = 1$. But this contradicts the fact that $m(\mathscr{C}) = 1$. Thus the first case cannot occur.

Now, assume we are in the second case. By the definition of μ_ϵ we have that

$$0 < \hat{m}(\mu_\epsilon^{-1}((0, \alpha])) \Rightarrow m(Q(\epsilon, \alpha)) \in (0, \alpha],$$

where $Q(\epsilon, \alpha) := \phi([0, \epsilon] \times \mu_\epsilon^{-1}((0, \alpha]))$. Take the union of the iterates of $Q(\epsilon, \alpha)$, that is

$$Q := \bigcup_{i \in \mathbb{Z}} f^i(Q(\epsilon, \alpha)),$$

which is an f-invariant set. Since $f_* m_y = m_{f(y)}$ then, for each $y \in Q$ we have

$$c(y) \cap Q \subset \{z : m_y([0_y, z]) \le \alpha\},$$

which implies $0 < m(Q) \le \alpha < 1$. This contradicts the ergodicity of f, thus the second case cannot occur as well. At last assume we are in the third case. In this case we have

$$\bigcup_{n \in \mathbb{N}} \hat{m}(\mu_\epsilon^{-1}((0, 1 - 1/n])) = 0 \Rightarrow \hat{m}(\mu_\epsilon^{-1}(\{1\})) = 1.$$

Let $\epsilon_0 := \inf\{\epsilon : \widehat{m}(\mu_\epsilon^{-1}(\{1\})) = 1\}$. Then, $\widehat{m}(\mu_{\epsilon_0+1/n}^{-1}(\{1\})) = 1$ for any $n \in \mathbb{N}$, which implies $m(Q(n)) = 1$, where $Q(n) := \phi([0, \epsilon_0 + 1/n] \times \mu_{\epsilon_0+1/n}^{-1}(\{1\}))$. Take $Q = \bigcap Q(n)$. We must have $m(Q) = 1$. But by the definition of $Q(n)$ and Q we have $Q = \phi([0, \epsilon_0] \times \mu_{\epsilon_0}^{-1}(\{1\}))$. Thus, by definition of ϵ_0 we have $m(\phi(\epsilon_0, \mu_{\epsilon_0}^{-1}(\{1\}))) = m(Q) = 1$, that is, the set of points $\Theta = \phi(\epsilon_0, \mu_{\epsilon_0}^{-1}(\{1\}))$ is a set of atoms as we wanted to show. Thus, the disintegration is indeed atomic, and there exists exactly one atom per leaf.   □

5.5 Proof of Theorem 3.6

Let $f : \mathbb{T}^3 \to \mathbb{T}^3$ be a C^2 volume preserving Kolmogorov partially hyperbolic diffeomorphism homotopic to a linear Anosov diffeomorphism $A : \mathbb{T}^3 \to \mathbb{T}^3$, as in the hypothesis of Theorem 3.6.

Without loss of generality we can assume that the center Lyapunov exponent λ_A^c of A is negative and $\mathscr{F}_f^{cs}$ is absolutely continuous: $\lambda_A^c < 0$ (otherwise we work with f^{-1} which is homotopic to A^{-1}). From the ergodicity of f it follows that $\mathscr{C}$ has either full or zero volume.

5.5.1 Partition by Rectangles

Take $\mathscr{E}$ to be the partition of $\mathbb{T}^3$ by points. Let α be a partition of $\mathbb{T}^3$ by measurable sets such that the boundary of any element in α is piecewise smooth and such that each atom $D \in \alpha$ is an open set with boundary of zero measure. It is easy to construct such a partition on the 3-torus. We will prove that α is VWB. Then we will take a sequence of such partitions α_n with: $\alpha_1 \leq \alpha_2 \leq \ldots$ such that $\alpha_n \to \mathscr{E}$, concluding that f is indeed Bernoulli.

Given two points y and z close enough to each other, we know that $\mathscr{F}^{cs}(y)$ and $\mathscr{F}^u(z)$ will intersect each other and that the intersection is, locally, a single point. We denote this point by $[y, z]$. Sometimes along this section we also write $W^{cs}(y) \cap W^u(z)$ to mean the point $[y, z]$.

Definition 5.15 (cf. Definitions 3.7 and 4.11) A measurable set Π is called a δ-rectangle at a point w if $\Pi \subset B(w, \delta)$ and for any $y, z \in \Pi$ the local intersection belongs to Π, that is $[y, z] \in \Pi$.

Note that, by the local product structure of the rectangles, we can think of a rectangle Π as a cartesian product of $\mathscr{F}_x^u \cap \Pi$ and $\mathscr{F}_z^{cs} \cap \Pi$, where $x, z \in \Pi$. Let $f : M \to M$ be a partially hyperbolic diffeomorphism with absolutely continuous center-stable foliation. Then we can take, for a typical z, m_z^u the measure m conditioned on $\mathscr{F}_z^u \cap \Pi$ and m_f^{cs} the factor measure on the leaf $\mathscr{F}_z^{cs}$. From the absolute continuity of the unstable and center-stable foliations it follows that for a typical z the product measure

$$m_R^P := m_z^u \times m_f^{cs},$$

which is defined on Π, satisfies $m_R^P << m$. Observe that the measure m_R^P depends on the point z, but notice that the absolute continuity with respect to m is independent of such typical point. Hence we will omit the choice of z in the forthcoming arguments unless it is strictly necessary.

The following definition of ε-regular covering was given by Chernov and Haskell [1] for the case of non-uniformly hyperbolic maps and flows. The definition for partially hyperbolic diffeomorphisms is the same replacing the stable by the center-stable manifold.

Definition 5.16 Given any $\varepsilon > 0$, an ε-regular covering of M is a finite collection of disjoint rectangles $\mathscr{R} = \mathscr{R}_\varepsilon$ such that:

1. $m(\bigcup_{R \in \mathscr{R}} R) > 1 - \varepsilon$
2. For every $R \in \mathscr{R}$ we have

$$\left| \frac{m_R^P(R)}{m(R)} - 1 \right| < \varepsilon$$

and, moreover, R contains a subset, G, with $m(G) > (1 - \varepsilon)m(R)$ which has the property that for all points in G,

$$\left| \frac{dm_R^P}{dm} - 1 \right| < \varepsilon.$$

Lemma 5.3 *Given any $\delta > 0$ and any $\varepsilon > 0$, there exists an ε-regular covering of connected rectangles $\mathscr{R}_\varepsilon$ of M with* $\mathrm{diam}(R) < \delta$*, for every $R \in \mathscr{R}_\varepsilon$.*

As remarked by Chernov and Haskell [1], in the non-uniformly hyperbolic case the expansion and contraction of the unstable and stable manifolds are not used to prove the existence of an ε-regular covering. The only properties used in the proof are the measurable dependence on $x \in M$, transversality between the unstable and stable foliations and the absolute continuity property.

By our hypothesis, all these properties are satisfied for the pair of foliations $\mathscr{F}^{cs}$ and $\mathscr{F}^u$. Now the proof of this lemma follows the same lines as the proof of Lemma 5.1 from [1] just replacing stable by center-stable foliation.

Definition 5.17 (cf. Definitions 3.8 and 4.13) We say that a measurable set A intersects a rectangle Π, *leafwise*, or that the intersection is a *u-tubular* subset, if

$$\mathscr{F}^u(w) \cap \Pi \subset A \cap \Pi, \text{ for any } w \in A \cap \Pi.$$

Lemma 5.4 *If E is a set intersecting a rectangle Π in a u-tubular subset, then the intersection $E \cap \Pi$ is a rectangle.*

Proof Let $x, y \in E \cap \Pi$ and set $z = W^u(x) \cap W^{cs}(y)$. Since $x, y \in \Pi$ and Π is a rectangle we have $z \in \Pi$. Now, because the intersection is leafwise we have $z \in \mathscr{F}^u(x) \cap \Pi \subset \mathscr{F}^u(x) \cap E$. $\square$

Lemma 5.5 *The set $\mathbb{T}^3 \setminus \mathscr{C}$ is u-saturated. In particular, given any rectangle Π, if $X = \mathbb{T}^3 \setminus \mathscr{C}$ then X intersects Π in a u-tubular subset.*

Proof Let $x \in \mathbb{T}^3 \setminus \mathscr{C}$ and assume that we can find $y \in \mathscr{F}^u(x) \cap \mathscr{C}$. Then, there exists a closed segment $\gamma \subset \mathscr{F}^c(y)$ with $y \in \gamma$ and $\gamma \subset \mathscr{C}$. Take any $w \in \gamma \setminus \{y\}$ and consider $z = W^u(w) \cap W^c(x)$.

Since we assumed that the center Lyapunov exponent of the linearization A of f is negative, the semi-conjugacy h sends strong unstable leaves of f to unstable leaves of A. Note that stable leaves of f are not necessarily mapped to stable leaves of A.

If $h(y) = h(w)$, then $\mathscr{F}^u(h(x)) = \mathscr{F}^u(h(y)) = \mathscr{F}^u(h(w)) = \mathscr{F}^u(h(z))$, which implies $h(z) \in \mathscr{F}^u(h(x)) \cap \mathscr{F}^c(h(x))$. Since $W^u(h(x)) \cap W^c(h(x)) = \{h(x)\}$ we can take z close enough to y and then we have $h(x) = h(z)$, contradicting the hypothesis that $x \notin \mathscr{C}$. $\square$

Lemma 5.6 *Given a rectangle Π and $\beta > 0$, one can find $N_1 > 0$ such that for any $N' \geq N \geq N_1$ and β-almost every element $A \in \bigvee_N^{N'} f^k \alpha$, there exists a subset $E \subset A$, intersecting Π leafwise, for which $m(E)/m(A) \geq 1 - \beta$.*

5.5.2   Construction of the Function θ

We now proceed to the most important part of the proof: the construction of the function θ satisfying the hypothesis of Lemma 3.2. The first step to do so is to obtain some control on the distance of orbits of points belonging to the same center-stable leaf.

Lemma 5.7 *Given any $\varepsilon > 0$ and any compact set $K \subset X$ in the essential injectivity domain of h (see Definition 5.14) there exists $n_0 \in \mathbb{N}$ such that for any two points $a \in K$, $b \in \mathscr{F}^{cs}(a) \cap K$, with $d(a, b) < \frac{1}{2}$ we have*

$$d(f^n(a), f^n(b)) < \varepsilon, \text{ whenever } f^n(a), f^n(b) \in K \text{ and } n \geq n_0.$$

Proof We split the proof in two cases.

First Case: $X = \mathbb{T}^3 \setminus \mathscr{C}$
Consider the lifts to the universal cover $\tilde{A}$, $\tilde{f} : \mathbb{R}^3 \to \mathbb{R}^3$ and $\tilde{h} : \mathbb{R}^3 \to \mathbb{R}^3$ the lift of the conjugacy, such that $\tilde{h}(0) = 0$. We know that $\tilde{A}^n \circ \tilde{h} = \tilde{h} \circ \tilde{f}^n$ for all n, thus

$$(e^{\lambda_A^c})^n d(\tilde{h}(a), \tilde{h}(b)) \geq d(\tilde{A}^n \circ \tilde{h}(a), \tilde{A}^n \circ \tilde{h}(b)) = d(\tilde{h} \circ \tilde{f}^n(a), \tilde{h} \circ \tilde{f}^n(b)).$$

Since h is a bounded distance from the identity, then if $d(a, b) < 1/2$ we have:

$$d(\tilde{A}^n \circ \tilde{h}(a), \tilde{A}^n \circ \tilde{h}(b)) \leq (e^{\lambda_A^c})^n \cdot D,$$

for a certain constant $D > 0$. By the uniform continuity of h^{-1} inside $h(K)$ we can take n_0 big enough so that $n \geq n_0$ implies $d(f^n(a), f^n(b)) < \varepsilon$.

Second Case: X = Set of Atoms

In this case, we know by Theorem 5.12 that each leaf has only one atom, that is, for almost every $x \in \mathbb{T}^3$ we have $X \cap \mathscr{F}^c(x) = \{a_x\}$. Thus, given any two points $a, b \in X$, a and b are not collapsed by h (since they do not belong to the same central leaf). Thus the proof of the first case works for this case as well.

□

Lemma 5.8 *For any $\delta > 0$, there exists $0 < \delta_1 < \delta$ with the following property. Let Π be a δ_1-rectangle and E a set intersecting Π leafwise. Then we can construct a bijective function $\theta : E \cap \Pi \to \Pi$ such that for every measurable set $F \subset E \cap \Pi$ we have*

$$\frac{m_\Pi^P(\theta(F))}{m_\Pi^P(\Pi)} = \frac{m_\Pi^P(F)}{m_\Pi^P(E \cap \Pi)},$$

and for every $x \in E \cap \Pi$

$$\theta(x) \in \mathscr{F}^{cs}(x).$$

Proof Since E intersects Π leafwise by Lemma 5.4 we know that $E \cap \Pi$ is a sub-rectangle, and since the center stable foliation is absolutely continuous the intersection $\mathscr{F}^{cs}(x) \cap E \cap \Pi$ has positive Lebesgue measure for almost every $x \in E \cap \Pi$. Let $z \in E \cap \Pi$ be any point such that $\mathscr{F}^{cs}(x) \cap E \cap \Pi$ has positive Lebesgue measure. Because $\mathscr{F}^{cs}(x) \cap E \cap \Pi$ and $\mathscr{F}^{cs}(x) \cap \Pi$ are both probability Lebesgue spaces (with the normalized measures) we can construct a bijection $\theta_0 : \mathscr{F}^{cs}(x) \cap E \cap \Pi \to \mathscr{F}^{cs}(x) \cap \Pi$ preserving the normalized measures, that is, for any measurable subset $J \subset \mathscr{F}^{cs}(x) \cap E \cap \Pi$ we have

$$\frac{m_f^{cs}(\theta_0(J))}{m_\Pi^P(\Pi)} = \frac{m_f^{cs}(J)}{m_\Pi^P(E \cap \Pi)}. \tag{5.6}$$

Now, given any $y \in E \cap \Pi$ we define (using that the intersection is leafwise and inside the rectangle) $\theta(y) \in \Pi$ by $\theta(y) := (\pi_{y,x}^u)^{-1} \circ \theta_0 \circ \pi_{y,x}^u(y)$.

This $\theta : E \cap \Pi \to \Pi$ is well defined and $\theta(y) \in \mathscr{F}^{cs}(y) \cap \Pi$. By the definition of m_Π^P, given a measurable set $F \subset E \cap \Pi$ we have

$$m_{\Pi}^P(F) = \int m_f^{cs}(\pi_{y,x}^u(\mathscr{F}^{cs}(y) \cap F))dm^u(y).$$

Thus,

$$m_{\Pi}^P(\theta(F)) = \int m_f^{cs}(\pi_{y,x}^u(\mathscr{F}^{cs}(y) \cap \theta(F)))dm^u(y)$$

$$= \int m_f^{cs}(\pi_{y,x}^u \circ \theta(\mathscr{F}^{cs}(y) \cap F))dm^u(y)$$

$$= \int m_f^{cs}(\theta_0 \circ \pi_{y,x}^u(\mathscr{F}^{cs}(y) \cap F))dm^u(y).$$

Substituting (5.6) we have

$$m_{\Pi}^P(\theta(F)) = \int m_f^{cs}(\theta_0 \circ \pi_{y,x}^u(\mathscr{F}^{cs}(y) \cap F))dm^u(y)$$

$$= \frac{m_{\Pi}^P(\Pi)}{m_{\Pi}^P(E \cap \Pi)} \int m_f^{cs}(\pi_{y,x}^u(\mathscr{F}^{cs}(y) \cap F))dm^u(y)$$

$$= \frac{m_{\Pi}^P(\Pi)}{m_{\Pi}^P(E \cap \Pi)} \cdot m_{\Pi}^P(F)$$

as we wanted to show.   □

The following Lemma concludes the proof of Theorem 3.6.

Lemma 5.9 *Let α be a finite partition with the property that each atom of α has piecewise smooth boundary. Then α is VWB.*

The proof of Lemma 5.9 closely follows from the arguments already used by Pesin [25] and Chernov and Haskell [1] with the technical difference that, by Lemma 5.8, the function θ, constructed from a tubular intersection to a rectangle containing this intersection, does not preserve stable manifolds as in [7, 25] and [1]. Instead, θ preserves center-stable manifolds. Points belonging to the same center-stable manifold do not have the property of getting exponentially close to each other as we iterate the dynamics, thus we cannot directly say that given $\varepsilon > 0$ a large set of pairs of points on the same center-stable manifold will asymptotically visit the same atoms.

To overcome this difficulty we use Theorem 5.12 which states that either $m(\mathscr{C}) = 0$ and then we have defined $X = \mathbb{T}^3 \setminus \mathscr{C}$, or $m(\mathscr{C}) = 1$ and then we can take a full measure set $X \subset \mathbb{T}^3$ intersecting almost every center-leaf in exactly one point. By Lemma 5.7, we can choose arbitrarily large compact subsets $K \subset X$ such that any pair of point $x, y \in K \subset X$ have the property that if $y \in \mathscr{F}^{cs}(x)$ then the distance between $f^n(x)$ and $f^n(y)$ is very small every time both of them visits the set K simultaneously. The point is that since we can take K arbitrarily large and f is ergodic, the set of natural numbers $\{n : f^n(x), f^n(y) \in K\}$ has arbitrarily large density, this will allow us to conclude that indeed for a large set of points $x \in X$ the Cesaro-means of Lemma 3.2 are indeed arbitrarily small. For a detailed proof of Lemma 5.9 we refer the reader to [28].

References

1. Anosov, D.V.: Ergodic properties of geodesic flows on closed Riemannian manifolds of negative curvature. Dokl. Akad. Nauk SSSR **151**, 1250–1252 (1963)
2. Anosov, D.V.: Geodesic flows on closed Riemann manifolds with negative curvature. In: Proceedings of the Steklov Institute of Mathematics, vol. 90 (1967). Translated from the Russian by S. Feder, American Mathematical Society, Providence, R.I. iv+235 (1969)
3. Aumann, R.J.: Measurable utility and the measurable choice theorem. La Decision, Editions du Centre National de la Reserche Scientifique, pp. 15–26 (1969)
4. Barreira, L., Yakov, P.: Nonuniform Hyperbolicity. Dynamics of Systems with Nonzero Lyapunov Exponents. Cambridge University Press, Cambridge (2007)
5. Bogachev, V.I.: Measure Theory, vol. I. Springer, Berlin (2007)
6. Bonatti, C., Díaz, L.J., Viana, M.: Dynamics Beyond Uniform Hyperbolicity. A Global Geometric and Probabilistic Perspective. Encyclopaedia of Mathematical Sciences, vol. 102. Mathematical Physics, III. Springer, Berlin (2005)
7. Brin, M., Pesin, Y.: Partially hyperbolic dynamical systems. Math. USSR-Izv. **8** 177–218 (1974)
8. Brin, M., Burago, D., Ivanov, D.: On partially hyperbolic diffeomorphisms of 3-manifolds with commutative fundamental group. In: Modern Dynamical Systems and Applications, pp. 307–312. Cambridge University Press, Cambridge (2004)
9. Buzzi, J., Fisher, T.: Entropic stability beyond partial hyperbolicity. J. Mod. Dyn. **7**, 527–552 (2013)
10. Chernov, N.I., Haskell, C.: Nonuniformly hyperbolic K-systems are Bernoulli. Ergodic Theory Dyn. Syst. **16**, 19–44 (1996)
11. do Carmo, M.: Riemannian Geometry. Birkhäuser, Basel (1992)
12. Franks, J.: Anosov diffeomorphisms. In: Global Analysis, Berkeley, CA, 1968. Proceedings of Symposia in Pure Mathematics, vol. XIV, pp. 61–93. American Mathematical Society, Providence (1970)
13. Gallavotti, G., Ornstein, D.S.: Billiards and Bernoulli schemes. Commun. Math. Phys. **38**, 83–101 (1974)
14. Hammerlindl, A., Ures, R.: Ergodicity and partial hyperbolicity on the 3-torus. Commun. Contemp. Math. **16**(4), 1350038, 22 (2014)
15. Hertz, F.R., Hertz, M.A.R., Ures, R.: A non-dynamically coherent example in $\mathbb{T}^3$. Ann. Inst. H. Poincaré Anal. Non Linéaire **33**, 1023–1032 (2016)
16. Kanigowski, A., Rodriguez Hertz, F., Vinhage, K.: On the non-equivalence of the Bernoulli and K properties in dimension four. J. Mod. Dyn. **13**, 221–250 (2018)
17. Katok, A., Hasselblatt, B.: Introduction to the Modern Theory of Dynamical Systems. Encyclopedia of Mathematics and its Applications, vol. 54, Cambridge University Press, Cambridge (1995)
18. Katok, A., Strelcyn, J.-M.: Invariant Manifolds, Entropy and Billiards; Smooth Maps with Singularities. Lecture Notes in Mathematics, vol. 1222. Springer, Berlin (1986)
19. Kubo, I., Murata, H.: Perturbed billiard systems II, Bernoulli properties. Nagoya Math. J. **81**, 1–25 (1981)
20. Ledrappier, F., Lima, Y., Sarig, O.: Ergodic properties of equilibrium measures for smooth three dimensional flows. **91**, 65–106 (2016)

21. Lindenstrauss, E., Scmidt, K.: Invariant sets and invariant measures of nonexpansive group automorphisms. Isr. J. Math. **144**, 29–60 (2004)
22. Manning, A.: There are no new Anosov diffeomorphisms on tori. Am. J. Math. **96**, 422–429 (1974)
23. Pesin, Y.: Families of invariant manifolds corresponding to nonzero characteristic exponents. Math. USSR Izv. **10**, 1261–1305 (1976)
24. Pesin, Y.: Characteristic Lyapunov exponents and smooth ergodic theory. Russian Math. Surv. **32**, 55–114 (1977)
25. Pesin, Y.: Geodesic flows on closed Riemannian manifolds without focal points. Math. USSR Izv. **11**, 1195–1228 (1977)
26. Pesin, Y.: A description of the π-partition of a diffeomorphism with an invariant measure. Math. Notes **22**, 506–515 (1977)
27. Ponce, G., Tahzibi, A., Varão, R.: Minimal yet measurable foliations. J. Mod. Dyn. **8**, 93–107 (2014)
28. Ponce, G., Tahzibi, A., Varão, R.: On the Bernoulli property for certain partially hyperbolic diffeomorphisms. Adv. Math. **329**, 329–360 (2018)
29. Pugh, C., Shub, M.: Ergodic attractors. Trans. Am. Math. Soc. **312**, 1–54 (1989)
30. Pugh, C., Shub, M.: Stably ergodic dynamical systems and partial hyperbolicity. J. Complex. **13**, 125–179 (1997)
31. Ratner, M.: Anosov flows with Gibbs measures are also Bernoullian. Isr. J. Math. **(17)**, 380–391 (1974)
32. Shub, M., Wilkinson, A.: Pathological foliations and removable zero exponents. Invent. Math. **139**, 495–508 (2000)
33. Ures, U.: Intrinsic ergodicity of partially hyperbolic diffeomorphisms with a hyperbolic linear part. Proc. Am. Math. Soc. **140**, 1973–1985 (2012)

Index

© The Author(s), under exclusive licence to Springer Nature Switzerland AG 2019
G. Ponce, R. Varão, *An Introduction to the Kolmogorov–Bernoulli Equivalence*,
SpringerBriefs in Mathematics, https://doi.org/10.1007/978-3-030-27390-3